CHAPMAN & HALL/CRC
Texts in Statistical Science Series

Analysis of Failure and Survival Data
Peter J. Smith

The Analysis and Interpretation of Multivariate Data for Social Scientists
David J. Bartholomew, Fiona Steele, Irini Moustaki, and Jane Galbraith

The Analysis of Time Series — An Introduction, Fifth Edition
C. Chatfield

Applied Bayesian Forecasting and Time Series Analysis
A. Pole, M. West, and J. Harrison

Applied Nonparametric Statistical Methods, Third Edition
P. Sprent and N.C. Smeeton

Applied Statistics — Principles and Examples
D.R. Cox and E.J. Snell

Bayesian Data Analysis
A. Gelman, J. Carlin, H. Stern, and D. Rubin

Beyond ANOVA — Basics of Applied Statistics
R.G. Miller, Jr.

Computer-Aided Multivariate Analysis, Third Edition
A.A. Afifi and V.A. Clark

A Course in Categorical Data Analysis
T. Leonard

A Course in Large Sample Theory
T.S. Ferguson

Data Driven Statistical Methods
P. Sprent

Decision Analysis — A Bayesian Approach
J.Q. Smith

Elementary Applications of Probability Theory, Second Edition
H.C. Tuckwell

Elements of Simulation
B.J.T. Morgan

Epidemiology — Study Design and Data Analysis
M. Woodward

Essential Statistics, Fourth Edition
D.A.G. Rees

A First Course in Linear Model Theory
Nalini Ravishanker and Dipak K. Dey

Interpreting Data — A First Course in Statistics
A.J.B. Anderson

An Introduction to Generalized Linear Models, Second Edition
A.J. Dobson

Introduction to Multivariate Analysis
C. Chatfield and A.J. Collins

Introduction to Optimization Methods and their Applications in Statistics
B.S. Everitt

Large Sample Methods in Statistics
P.K. Sen and J. da Motta Singer

Markov Chain Monte Carlo — Stochastic Simulation for Bayesian Inference
D. Gamerman

Mathematical Statistics
K. Knight

Modeling and Analysis of Stochastic Systems
V. Kulkarni

Modelling Binary Data
D. Collett

Modelling Survival Data in Medical Research
D. Collett

Multivariate Analysis of Variance and Repeated Measures — A Practical Approach for Behavioural Scientists
D.J. Hand and C.C. Taylor

Multivariate Statistics — A Practical Approach
B. Flury and H. Riedwyl

Practical Data Analysis for Designed Experiments
B.S. Yandell

Practical Longitudinal Data Analysis
D.J. Hand and M. Crowder

Practical Statistics for Medical Research
D.G. Altman

Probability — Methods and Measurement
A. O'Hagan

Problem Solving — A Statistician's Guide, Second Edition
C. Chatfield

Randomization, Bootstrap and Monte Carlo Methods in Biology, Second Edition
B.F.J. Manly

Readings in Decision Analysis
S. French

Sampling Methodologies with Applications
P. Rao

Statistical Analysis of Reliability Data
M.J. Crowder, A.C. Kimber, T.J. Sweeting and R.L. Smith

Statistical Methods for SPC and TQM
D. Bissell

Statistical Methods in Agriculture and Experimental Biology, Second Edition
R. Mead, R.N. Curnow and A.M. Hasted

Statistical Process Control — Theory and Practice, Third Edition
G.B. Wetherill and D.W. Brown

Statistical Theory, Fourth Edition
B.W. Lindgren

Statistics for Accountants, Fourth Edition
S. Letchford

Statistics for Technology — A Course in Applied Statistics, Third Edition
C. Chatfield

Statistics in Engineering — A Practical Approach
A.V. Metcalfe

Statistics in Research and Development, Second Edition
R. Caulcutt

The Theory of Linear Models
B. Jørgensen

AN INTRODUCTION TO GENERALIZED LINEAR MODELS

SECOND EDITION

Annette J. Dobson

CHAPMAN & HALL/CRC

A CRC Press Company

Boca Raton London New York Washington, D.C.

Library of Congress Cataloging-in-Publication Data

Dobson, Annette J., 1945-
An introduction to generalized linear models / Annette J. Dobson.—2nd ed.
p. cm.— (Chapman & Hall/CRC texts in statistical science series)
Includes bibliographical references and index.
ISBN 1-58488-165-8 (alk. paper)
1. Linear models (Statistics) I. Title. II. Texts in statistical science.

QA276 .D589 2001
519.5′35—dc21 2001047417

Direct all inquiries to CRC Press LLC, 2000 N.W. Corporate Blvd., Boca Raton, Florida 33431.

Visit the CRC Press Web site at www.crcpress.com

International Standard Book Number 1-58488-165-8
Library of Congress Card Number 2001047417
Printed in the United States of America 1 2 3 4 5 6 7 8 9 0
Printed on acid-free paper

Contents

Preface

Statistical tools for analyzing data are developing rapidly so that the 1990 edition of this book is now out of date.

The original purpose of the book was to present a unified theoretical and conceptual framework for statistical modelling in a way that was accessible to undergraduate students and researchers in other fields. This new edition has been expanded to include nominal (or multinomial) and ordinal logistic regression, survival analysis and analysis of longitudinal and clustered data. Although these topics do not fall strictly within the definition of generalized linear models, the underlying principles and methods are very similar and their inclusion is consistent with the original purpose of the book.

The new edition relies on numerical methods more than the previous edition did. Some of the calculations can be performed with a spreadsheet while others require statistical software. There is an emphasis on graphical methods for exploratory data analysis, visualizing numerical optimization (for example, of the likelihood function) and plotting residuals to check the adequacy of models.

The data sets and outline solutions of the exercises are available on the publisher's website:
www.crcpress.com/us/ElectronicProducts/downandup.asp?mscssid=

I am grateful to colleagues and students at the Universities of Queensland and Newcastle, Australia, for their helpful suggestions and comments about the material.

Annette Dobson

1

Introduction

1.1 Background

This book is designed to introduce the reader to generalized linear models; these provide a unifying framework for many commonly used statistical techniques. They also illustrate the ideas of statistical modelling.

The reader is assumed to have some familiarity with statistical principles and methods. In particular, understanding the concepts of estimation, sampling distributions and hypothesis testing is necessary. Experience in the use of t-tests, analysis of variance, simple linear regression and chi-squared tests of independence for two-dimensional contingency tables is assumed. In addition, some knowledge of matrix algebra and calculus is required.

The reader will find it necessary to have access to statistical computing facilities. Many statistical programs, languages or packages can now perform the analyses discussed in this book. Often, however, they do so with a different program or procedure for each type of analysis so that the unifying structure is not apparent.

Some programs or languages which have procedures consistent with the approach used in this book are: **Stata, S-PLUS, Glim**, **Genstat** and **SYSTAT**. This list is not comprehensive as appropriate modules are continually being added to other programs.

In addition, anyone working through this book may find it helpful to be able to use mathematical software that can perform matrix algebra, differentiation and iterative calculations.

1.2 Scope

The statistical methods considered in this book all involve the analysis of relationships between measurements made on groups of subjects or objects. For example, the measurements might be the heights or weights and the ages of boys and girls, or the yield of plants under various growing conditions. We use the terms **response**, **outcome** or **dependent variable** for measurements that are free to vary in response to other variables called **explanatory variables** or **predictor variables** or **independent variables** - although this last term can sometimes be misleading. Responses are regarded as random variables. Explanatory variables are usually treated as though they are non-random measurements or observations; for example, they may be fixed by the experimental design.

Responses and explanatory variables are measured on one of the following scales.

1. **Nominal** classifications: e.g., red, green, blue; yes, no, do not know, not applicable. In particular, for **binary, dichotomous** or **binomial** variables

there are only two categories: male, female; dead, alive; smooth leaves, serrated leaves. If there are more than two categories the variable is called **polychotomous, polytomous** or **multinomial**.

2. **Ordinal** classifications in which there is some natural order or ranking between the categories: e.g., young, middle aged, old; diastolic blood pressures grouped as ≤ 70, 71-90, 91-110, 111-130, $\geq$131mm Hg.

3. **Continuous** measurements where observations may, at least in theory, fall anywhere on a continuum: e.g., weight, length or time. This scale includes both **interval scale** and **ratio scale** measurements – the latter have a well-defined zero. A particular example of a continuous measurement is the time until a specific event occurs, such as the failure of an electronic component; the length of time from a known starting point is called the **failure time**.

Nominal and ordinal data are sometimes called **categorical** or **discrete variables** and the numbers of observations, **counts** or **frequencies** in each category are usually recorded. For continuous data the individual measurements are recorded. The term **quantitative** is often used for a variable measured on a continuous scale and the term **qualitative** for nominal and sometimes for ordinal measurements. A qualitative, explanatory variable is called a **factor** and its categories are called the **levels** for the factor. A quantitative explanatory variable is sometimes called a **covariate**.

Methods of statistical analysis depend on the measurement scales of the response and explanatory variables.

This book is mainly concerned with those statistical methods which are relevant when there is just *one response variable,* although there will usually be several explanatory variables. The responses measured on different subjects are usually assumed to be statistically independent random variables although this requirement is dropped in the final chapter which is about correlated data. Table 1.1 shows the main methods of statistical analysis for various combinations of response and explanatory variables and the chapters in which these are described.

The present chapter summarizes some of the statistical theory used throughout the book. Chapters 2 to 5 cover the theoretical framework that is common to the subsequent chapters. Later chapters focus on methods for analyzing particular kinds of data.

Chapter 2 develops the main ideas of statistical modelling. The modelling process involves four steps:

1. Specifying models in two parts: equations linking the response and explanatory variables, and the probability distribution of the response variable.

2. Estimating parameters used in the models.

3. Checking how well the models fit the actual data.

4. Making inferences; for example, calculating confidence intervals and testing hypotheses about the parameters.

Table 1.1 *Major methods of statistical analysis for response and explanatory variables measured on various scales and chapter references for this book.*

Response (chapter)	Explanatory variables	Methods
Continuous (Chapter 6)	Binary	t-test
	Nominal, >2 categories	Analysis of variance
	Ordinal	Analysis of variance
	Continuous	Multiple regression
	Nominal & some continuous	Analysis of covariance
	Categorical & continuous	Multiple regression
Binary (Chapter 7)	Categorical	Contingency tables Logistic regression
	Continuous	Logistic, probit & other dose-response models
	Categorical & continuous	Logistic regression
Nominal with >2 categories (Chapter 8 & 9)	Nominal	Contingency tables
	Categorical & continuous	Nominal logistic regression
Ordinal (Chapter 8)	Categorical & continuous	Ordinal logistic regression
Counts (Chapter 9)	Categorical	Log-linear models
	Categorical & continuous	Poisson regression
Failure times (Chapter 10)	Categorical & continuous	Survival analysis (parametric)
Correlated responses (Chapter 11)	Categorical & continuous	Generalized estimating equations Multilevel models

The next three chapters provide the theoretical background. Chapter 3 is about the **exponential family of distributions**, which includes the Normal, Poisson and binomial distributions. It also covers **generalized linear models** (as defined by Nelder and Wedderburn, 1972). Linear regression and many other models are special cases of generalized linear models. In Chapter 4 methods of estimation and model fitting are described.

Chapter 5 outlines methods of statistical inference for generalized linear models. Most of these are based on how well a model describes the set of data. For example, **hypothesis testing** is carried out by first specifying alternative models (one corresponding to the null hypothesis and the other to a more general hypothesis). Then test statistics are calculated which measure the 'goodness of fit' of each model and these are compared. Typically the model corresponding to the null hypothesis is simpler, so if it fits the data about as well as a more complex model it is usually preferred on the grounds of parsimony (i.e., we retain the null hypothesis).

Chapter 6 is about **multiple linear regression** and **analysis of variance** (ANOVA). Regression is the standard method for relating a continuous response variable to several continuous explanatory (or predictor) variables. ANOVA is used for a continuous response variable and categorical or qualitative explanatory variables (factors). **Analysis of covariance** (ANCOVA) is used when at least one of the explanatory variables is continuous. Nowadays it is common to use the same computational tools for all such situations. The terms **multiple regression** or **general linear model** are used to cover the range of methods for analyzing one continuous response variable and multiple explanatory variables.

Chapter 7 is about methods for analyzing binary response data. The most common one is **logistic regression** which is used to model relationships between the response variable and several explanatory variables which may be categorical or continuous. Methods for relating the response to a single continuous variable, the dose, are also considered; these include **probit analysis** which was originally developed for analyzing dose-response data from bioassays. Logistic regression has been generalized in recent years to include responses with more than two nominal categories (nominal, **multinomial**, **polytomous** or **polychotomous logistic regression**) or ordinal categories (**ordinal logistic regression**). These new methods are discussed in Chapter 8.

Chapter 9 concerns **count** data. The counts may be frequencies displayed in a **contingency table** or numbers of events, such as traffic accidents, which need to be analyzed in relation to some 'exposure' variable such as the number of motor vehicles registered or the distances travelled by the drivers. Modelling methods are based on assuming that the distribution of counts can be described by the Poisson distribution, at least approximately. These methods include **Poisson regression** and **log-linear models.**

Survival analysis is the usual term for methods of analyzing failure time data. The parametric methods described in Chapter 10 fit into the framework of generalized linear models although the probability distribution assumed for the failure times may not belong to the exponential family.

Generalized linear models have been extended to situations where the responses are correlated rather than independent random variables. This may occur, for instance, if they are **repeated measurements** on the same subject or measurements on a group of related subjects obtained, for example, from **clustered sampling**. The method of **generalized estimating equations** (GEE's) has been developed for analyzing such data using techniques analogous to those for generalized linear models. This method is outlined in Chapter 11 together with a different approach to correlated data, namely **multilevel modelling.**

Further examples of generalized linear models are discussed in the books by McCullagh and Nelder (1989), Aitkin et al. (1989) and Healy (1988). Also there are many books about specific generalized linear models such as Hosmer and Lemeshow (2000), Agresti (1990, 1996), Collett (1991, 1994), Diggle, Liang and Zeger (1994), and Goldstein (1995).

1.3 Notation

Generally we follow the convention of denoting random variables by upper case italic letters and observed values by the corresponding lower case letters. For example, the observations $y_1, y_2, ..., y_n$ are regarded as realizations of the random variables $Y_1, Y_2, ..., Y_n$. Greek letters are used to denote parameters and the corresponding lower case roman letters are used to denote estimators and estimates; occasionally the symbol $\widehat{}$ is used for estimators or estimates. For example, the parameter β is estimated by $\widehat{\beta}$ or b. Sometimes these conventions are not strictly adhered to, either to avoid excessive notation in cases where the meaning should be apparent from the context, or when there is a strong tradition of alternative notation (e.g., e or ε for random error terms).

Vectors and matrices, whether random or not, are denoted by bold face lower and upper case letters, respectively. Thus, $\mathbf{y}$ represents a vector of observations

$$\begin{bmatrix} y_1 \\ \vdots \\ y_n \end{bmatrix}$$

or a vector of random variables

$$\begin{bmatrix} Y_1 \\ \vdots \\ Y_n \end{bmatrix},$$

$\boldsymbol{\beta}$ denotes a vector of parameters and $\mathbf{X}$ is a matrix. The superscript ^{T} is used for a matrix transpose or when a column vector is written as a row, e.g., $\mathbf{y} = [Y_1, ..., Y_n]^T$.

The probability density function of a continuous random variable Y (or the probability mass function if Y is discrete) is referred to simply as a **probability distribution** and denoted by

$$f(y; \boldsymbol{\theta})$$

where $\boldsymbol{\theta}$ represents the parameters of the distribution.

We use dot $(\cdot)$ subscripts for summation and bars $(^{-})$ for means, thus

$$\overline{y} = \frac{1}{N}\sum_{i=1}^{N} y_i = \frac{1}{N}\, y\cdot .$$

The expected value and variance of a random variable Y are denoted by $\mathrm{E}(Y)$ and $\mathrm{var}(Y)$ respectively. Suppose random variables $Y_1, ..., Y_N$ are independent with $\mathrm{E}(Y_i) = \mu_i$ and $\mathrm{var}(Y_i) = \sigma_i^2$ for $i = 1, ..., n$. Let the random variable W be a **linear combination** of the Y_i's

$$W = a_1Y_1 + a_2Y_2 + ... + a_nY_n, \tag{1.1}$$

where the a_i's are constants. Then the expected value of W is

$$\mathrm{E}(W) = a_1\mu_1 + a_2\mu_2 + ... + a_n\mu_n \tag{1.2}$$

and its variance is

$$\mathrm{var}(W) = a_1^2\sigma_1^2 + a_2^2\sigma_2^2 + ... + a_n^2\sigma_n^2. \tag{1.3}$$

1.4 Distributions related to the Normal distribution

The sampling distributions of many of the estimators and test statistics used in this book depend on the Normal distribution. They do so either directly because they are derived from Normally distributed random variables, or asymptotically, via the Central Limit Theorem for large samples. In this section we give definitions and notation for these distributions and summarize the relationships between them. The exercises at the end of the chapter provide practice in using these results which are employed extensively in subsequent chapters.

1.4.1 Normal distributions

1. If the random variable Y has the Normal distribution with mean μ and variance σ^2, its probability density function is

$$f(y; \mu, \sigma^2) = \frac{1}{\sqrt{2\pi\sigma^2}} \exp\left[-\frac{1}{2}\left(\frac{y-\mu}{\sigma^2}\right)^2\right].$$

   We denote this by $Y \sim N(\mu, \sigma^2)$.

2. The Normal distribution with $\mu = 0$ and $\sigma^2 = 1$, $Y \sim N(0, 1)$, is called the **standard Normal distribution**.

3. Let $Y_1, ..., Y_n$ denote Normally distributed random variables with $Y_i \sim N(\mu_i, \sigma_i^2)$ for $i = 1, ..., n$ and let the covariance of Y_i and Y_j be denoted by

$$\text{cov}(Y_i, Y_j) = \rho_{ij}\sigma_i\sigma_j \ ,$$

where ρ_{ij} is the correlation coefficient for Y_i and Y_j. Then the joint distribution of the Y_i's is the **multivariate Normal distribution** with mean vector $\boldsymbol{\mu} = [\mu_1, ..., \mu_n]^T$ and variance-covariance matrix $\mathbf{V}$ with diagonal elements σ_i^2 and non-diagonal elements $\rho_{ij}\sigma_i\sigma_j$ for $i \neq j$. We write this as $\mathbf{y} \sim \mathbf{N}(\boldsymbol{\mu}, \mathbf{V})$, where $\mathbf{y} = [Y_1, ..., Y_n]^T$.

4. Suppose the random variables $Y_1, ..., Y_n$ are independent and Normally distributed with the distributions $Y_i \sim N(\mu_i, \sigma_i^2)$ for $i = 1, ..., n$. If

$$W = a_1Y_1 + a_2Y_2 + ... + a_nY_n,$$

where the a_i's are constants. Then W is also Normally distributed, so that

$$W = \sum_{i=1}^{n} a_iY_i \sim N\left(\sum_{i=1}^{n} a_i\mu_i, \ \sum_{i=1}^{n} a_i^2\sigma_i^2\right)$$

by equations (1.2) and (1.3).

1.4.2 Chi-squared distribution

1. The **central chi-squared distribution** with n degrees of freedom is defined as the sum of squares of n independent random variables $Z_1, ..., Z_n$ each with the standard Normal distribution. It is denoted by

$$X^2 = \sum_{i=1}^{n} Z_i^2 \sim \chi^2(n).$$

In matrix notation, if $\mathbf{z} = [Z_1, ..., Z_n]^T$ then $\mathbf{z}^T\mathbf{z} = \sum_{i=1}^n Z_i^2$ so that $X^2 = \mathbf{z}^T\mathbf{z} \sim \chi^2(n)$.

2. If X^2 has the distribution $\chi^2(n)$, then its expected value is $\text{E}(X^2) = n$ and its variance is $\text{var}(X^2) = 2n$.

3. If $Y_1, ..., Y_n$ are independent Normally distributed random variables each with the distribution $Y_i \sim N(\mu_i, \sigma_i^2)$ then

$$X^2 = \sum_{i=1}^{n} \left(\frac{Y_i - \mu_i}{\sigma_i}\right)^2 \sim \chi^2(n) \tag{1.4}$$

because each of the variables $Z_i = (Y_i - \mu_i)/\sigma_i$ has the standard Normal distribution $N(0, 1)$.

4. Let $Z_1, ..., Z_n$ be independent random variables each with the distribution $N(0, 1)$ and let $Y_i = Z_i + \mu_i$, where at least one of the μ_i's is non-zero. Then the distribution of

$$\sum Y_i^2 = \sum (Z_i + \mu_i)^2 = \sum Z_i^2 + 2\sum Z_i\mu_i + \sum \mu_i^2$$

has larger mean $n + \lambda$ and larger variance $2n + 4\lambda$ than $\chi^2(n)$ where $\lambda = \sum \mu_i^2$. This is called the **non-central chi-squared distribution** with n degrees of freedom and **non-centrality parameter** λ. It is denoted by $\chi^2(n, \lambda)$.

5. Suppose that the Y_i's are not necessarily independent and the vector $\mathbf{y} = [Y_1, \ldots, Y_n]^T$ has the multivariate normal distribution $\mathbf{y} \sim \mathbf{N}(\boldsymbol{\mu}, \mathbf{V})$ where the variance-covariance matrix $\mathbf{V}$ is non-singular and its inverse is $\mathbf{V}^{-1}$. Then

$$X^2 = (\mathbf{y} - \boldsymbol{\mu})^T \mathbf{V}^{-1} (\mathbf{y} - \boldsymbol{\mu}) \sim \chi^2(n). \tag{1.5}$$

6. More generally if $\mathbf{y} \sim \mathbf{N}(\boldsymbol{\mu}, \mathbf{V})$ then the random variable $\mathbf{y}^T \mathbf{V}^{-1} \mathbf{y}$ has the non-central chi-squared distribution $\chi^2(n, \lambda)$ where $\lambda = \boldsymbol{\mu}^T \mathbf{V}^{-1} \boldsymbol{\mu}$.
7. If $X_1^2, \ldots, X_m^2$ are m independent random variables with the chi-squared distributions $X_i^2 \sim \chi^2(n_i, \lambda_i)$, which may or may not be central, then their sum also has a chi-squared distribution with $\sum n_i$ degrees of freedom and non-centrality parameter $\sum \lambda_i$, i.e.,

$$\sum_{i=1}^{m} X_i^2 \sim \chi^2 \left(\sum_{i=1}^{m} n_i, \sum_{i=1}^{m} \lambda_i \right).$$

   This is called the **reproductive property** of the chi-squared distribution.
8. Let $\mathbf{y} \sim \mathbf{N}(\boldsymbol{\mu}, \mathbf{V})$, where $\mathbf{y}$ has n elements but the Y_i's are not independent so that $\mathbf{V}$ is singular with rank $k < n$ and the inverse of $\mathbf{V}$ is not uniquely defined. Let $\mathbf{V}^-$denote a generalized inverse of $\mathbf{V}$. Then the random variable $\mathbf{y}^T \mathbf{V}^- \mathbf{y}$ has the non-central chi-squared distribution with k degrees of freedom and non-centrality parameter $\lambda = \boldsymbol{\mu}^T \mathbf{V}^- \boldsymbol{\mu}$.

   For further details about properties of the chi-squared distribution see Rao (1973, Chapter 3).

1.4.3 t-distribution

The **t-distribution** with n degrees of freedom is defined as the ratio of two independent random variables. The numerator has the standard Normal distribution and the denominator is the square root of a central chi-squared random variable divided by its degrees of freedom; that is,

$$T = \frac{Z}{(X^2/n)^{1/2}} \tag{1.6}$$

where $Z \sim N(0, 1)$, $X^2 \sim \chi^2(n)$ and Z and X^2 are independent. This is denoted by $T \sim t(n)$.

1.4.4 F-distribution

1. The **central F-distribution** with n and m degrees of freedom is defined as the ratio of two independent central chi-squared random variables each

divided by its degrees of freedom,

$$F = \frac{X_1^2}{n} / \frac{X_2^2}{m} \tag{1.7}$$

where $X_1^2 \sim \chi^2(n), X_2^2 \sim \chi^2(m)$ and X_1^2 and X_2^2 are independent. This is denoted by $F \sim F(n, m)$.

2. The relationship between the t-distribution and the F-distribution can be derived by squaring the terms in equation (1.6) and using definition (1.7) to obtain

$$T^2 = \frac{Z^2}{1} / \frac{X^2}{n} \sim F(1, n), \tag{1.8}$$

that is, the square of a random variable with the t-distribution, $t(n)$, has the F-distribution, $F(1, n)$.

3. The **non-central F-distribution** is defined as the ratio of two independent random variables, each divided by its degrees of freedom, where the numerator has a non-central chi-squared distribution and the denominator has a central chi-squared distribution, i.e.,

$$F = \frac{X_1^2}{n} / \frac{X_2^2}{m}$$

where $X_1^2 \sim \chi^2(n, \lambda)$ with $\lambda = \boldsymbol{\mu}^T \mathbf{V}^{-1} \boldsymbol{\mu}$, $X_2^2 \sim \chi^2(m)$ and X_1^2 and X_2^2 are independent. The mean of a non-central F-distribution is larger than the mean of central F-distribution with the same degrees of freedom.

1.5 Quadratic forms

1. A **quadratic form** is a polynomial expression in which each term has degree 2. Thus $y_1^2 + y_2^2$ and $2y_1^2 + y_2^2 + 3y_1y_2$ are quadratic forms in y_1 and y_2 but $y_1^2 + y_2^2 + 2y_1$ or $y_1^2 + 3y_2^2 + 2$ are not.

2. Let $\mathbf{A}$ be a symmetric matrix

$$\begin{bmatrix} a_{11} & a_{12} & \cdots & a_{1n} \\ a_{21} & a_{22} & \cdots & a_{2n} \\ \vdots & & \ddots & \vdots \\ a_{n1} & a_{n2} & \cdots & a_{nn} \end{bmatrix},$$

where $a_{ij} = a_{ji}$, then the expression $\mathbf{y}^T \mathbf{A} \mathbf{y} = \sum_i \sum_j a_{ij} y_i y_j$ is a quadratic form in the y_i's. The expression $(\mathbf{y} - \boldsymbol{\mu})^T \mathbf{V}^{-1} (\mathbf{y} - \boldsymbol{\mu})$ is a quadratic form in the terms $(y_i - \mu_i)$ but not in the y_i's.

3. The quadratic form $\mathbf{y}^T \mathbf{A} \mathbf{y}$ and the matrix $\mathbf{A}$ are said to be **positive definite** if $\mathbf{y}^T \mathbf{A} \mathbf{y} > 0$ whenever the elements of $\mathbf{y}$ are not all zero. A necessary and sufficient condition for positive definiteness is that all the determinants

$$|A_1| = a_{11}, |A_2| = \begin{vmatrix} a_{11} & a_{12} \\ a_{21} & a_{22} \end{vmatrix}, |A_3| = \begin{vmatrix} a_{11} & a_{12} & a_{13} \\ a_{21} & a_{22} & a_{23} \\ a_{31} & a_{32} & a_{33} \end{vmatrix}, \ldots, \text{ and}$$

$|A_n| = \det \mathbf{A}$ are all positive.

4. The rank of the matrix $\mathbf{A}$ is also called the degrees of freedom of the quadratic form $Q = \mathbf{y}^T \mathbf{A} \mathbf{y}$.

5. Suppose $Y_1, ..., Y_n$ are independent random variables each with the Normal distribution $N(0, \sigma^2)$. Let $Q = \sum_{i=1}^{n} Y_i^2$ and let $Q_1, ..., Q_k$ be quadratic forms in the Y_i's such that

$$Q = Q_1 + ... + Q_k$$

where Q_i has m_i degrees of freedom ($i = 1, \ldots, k$). Then

$Q_1, ..., Q_k$ are independent random variables and

$Q_1/\sigma^2 \sim \chi^2(m_1)$, $Q_2/\sigma^2 \sim \chi^2(m_2)$, $\cdots$ and $Q_k/\sigma^2 \sim \chi^2(m_k)$,

if and only if,

$$m_1 + m_2 + ... + m_k = n.$$

This is Cochran's theorem; for a proof see, for example, Hogg and Craig (1995). A similar result holds for non-central distributions; see Chapter 3 of Rao (1973).

6. A consequence of Cochran's theorem is that the difference of two independent random variables, $X_1^2 \sim \chi^2(m)$ and $X_2^2 \sim \chi^2(k)$, also has a chi-squared distribution

$$X^2 = X_1^2 - X_2^2 \sim \chi^2(m - k)$$

provided that $X^2 \geq 0$ and $m > k$.

1.6 Estimation

1.6.1 Maximum likelihood estimation

Let $\mathbf{y} = [Y_1, ..., Y_n]^T$ denote a random vector and let the joint probability density function of the Y_i's be

$$f(\mathbf{y}; \boldsymbol{\theta})$$

which depends on the vector of parameters $\boldsymbol{\theta} = [\theta_1, ..., \theta_p]^T$.

The **likelihood function** $L(\boldsymbol{\theta}; \mathbf{y})$ is algebraically the same as the joint probability density function $f(\mathbf{y}; \boldsymbol{\theta})$ but the change in notation reflects a shift of emphasis from the random variables $\mathbf{y}$, with $\boldsymbol{\theta}$ fixed, to the parameters $\boldsymbol{\theta}$ with $\mathbf{y}$ fixed. Since L is defined in terms of the random vector $\mathbf{y}$, it is itself a random variable. Let Ω denote the set of all possible values of the parameter vector $\boldsymbol{\theta}$; Ω is called the **parameter space**. The **maximum likelihood estimator** of $\boldsymbol{\theta}$ is the value $\widehat{\boldsymbol{\theta}}$ which maximizes the likelihood function, that is

$$L(\widehat{\boldsymbol{\theta}}; \mathbf{y}) \geq L(\boldsymbol{\theta}; \mathbf{y}) \qquad \text{for all } \boldsymbol{\theta} \text{ in } \Omega.$$

Equivalently, $\widehat{\boldsymbol{\theta}}$ is the value which maximizes the **log-likelihood function**

$l(\boldsymbol{\theta}; \mathbf{y}) = \log L(\boldsymbol{\theta}; \mathbf{y})$, since the logarithmic function is monotonic. Thus

$$l(\widehat{\boldsymbol{\theta}}; \mathbf{y}) \geq l(\boldsymbol{\theta}; \mathbf{y}) \qquad \text{for all } \boldsymbol{\theta} \text{ in } \Omega.$$

Often it is easier to work with the log-likelihood function than with the likelihood function itself.

Usually the estimator $\widehat{\boldsymbol{\theta}}$ is obtained by differentiating the log-likelihood function with respect to each element θ_j of $\boldsymbol{\theta}$ and solving the simultaneous equations

$$\frac{\partial l(\boldsymbol{\theta}; \mathbf{y})}{\partial \theta_j} = 0 \qquad \text{for } j = 1, ..., p. \tag{1.9}$$

It is necessary to check that the solutions do correspond to maxima of $l(\boldsymbol{\theta}; \mathbf{y})$ by verifying that the matrix of second derivatives

$$\frac{\partial^2 l(\boldsymbol{\theta}; \mathbf{y})}{\partial \theta_j \partial \theta_k}$$

evaluated at $\boldsymbol{\theta} = \widehat{\boldsymbol{\theta}}$ is negative definite. For example, if $\boldsymbol{\theta}$ has only one element θ this means it is necessary to check that

$$\left[\frac{\partial^2 l(\theta, \ \mathbf{y})}{\partial \theta^2}\right]_{\theta=\widehat{\theta}} < 0.$$

It is also necessary to check if there are any values of $\boldsymbol{\theta}$ at the edges of the parameter space Ω that give local maxima of $l(\boldsymbol{\theta}; \mathbf{y})$. When all local maxima have been identified, the value of $\widehat{\boldsymbol{\theta}}$ corresponding to the largest one is the maximum likelihood estimator. (For most of the models considered in this book there is only one maximum and it corresponds to the solution of the equations $\partial l / \partial \theta_j = 0, j = 1, ..., p$.)

An important property of maximum likelihood estimators is that if $g(\boldsymbol{\theta})$ is any function of the parameters $\boldsymbol{\theta}$, then the maximum likelihood estimator of $g(\boldsymbol{\theta})$ is $g(\widehat{\boldsymbol{\theta}})$. This follows from the definition of $\widehat{\boldsymbol{\theta}}$. It is sometimes called the **invariance property** of maximum likelihood estimators. A consequence is that we can work with a function of the parameters that is convenient for maximum likelihood estimation and then use the invariance property to obtain maximum likelihood estimates for the required parameters.

In principle, it is not necessary to be able to find the derivatives of the likelihood or log-likelihood functions or to solve equation (1.9) if $\widehat{\boldsymbol{\theta}}$ can be found numerically. In practice, numerical approximations are very important for generalized linear models.

Other properties of maximum likelihood estimators include consistency, sufficiency, asymptotic efficiency and asymptotic normality. These are discussed in books such as Cox and Hinkley (1974) or Kalbfleisch (1985, Chapters 1 and 2).

1.6.2 Example: Poisson distribution

Let $Y_1, ..., Y_n$ be independent random variables each with the Poisson distribution

$$f(y_i; \theta) = \frac{\theta^{y_i} e^{-\theta}}{y_i!}, \qquad y_i = 0, 1, 2, ...$$

with the same parameter θ. Their joint distribution is

$$\begin{aligned} f(y_1, \ldots, y_n; \theta) &= \prod_{i=1}^{n} f(y_i; \theta) = \frac{\theta^{y_1} e^{-\theta}}{y_1!} \times \frac{\theta^{y_2} e^{-\theta}}{y_2!} \times \cdots \times \frac{\theta^{y_n} e^{-\theta}}{y_n!} \\ &= \frac{\theta^{\sum y_i} \; e^{-n\theta}}{y_1! y_2! ... y_n!}. \end{aligned}$$

This is also the likelihood function $L(\theta; y_1, ..., y_n)$. It is easier to use the log-likelihood function

$$l(\theta; y_1, ..., y_n) = \log L(\theta; y_1, ..., y_n) = (\sum y_i) \log \theta - n\theta - \sum (\log y_i!).$$

To find the maximum likelihood estimate $\widehat{\theta}$, use

$$\frac{dl}{d\theta} = \frac{1}{\theta} \sum y_i - n.$$

Equate this to zero to obtain the solution

$$\widehat{\theta} = \sum y_i / n = \overline{y}.$$

Since $d^2 l / d\theta^2 = -\sum y_i / \theta^2 < 0$, l has its maximum value when $\theta = \widehat{\theta}$, confirming that $\overline{y}$ is the maximum likelihood estimate.

1.6.3 Least Squares Estimation

Let $Y_1, ..., Y_n$ be independent random variables with expected values $\mu_1, ..., \mu_n$ respectively. Suppose that the μ_i's are functions of the parameter vector that we want to estimate, $\boldsymbol{\beta} = [\beta_1, ..., \beta_p]^T$, $p < n$. Thus

$$E(Y_i) = \mu_i(\boldsymbol{\beta}).$$

The simplest form of the **method of least squares** consists of finding the estimator $\widehat{\boldsymbol{\beta}}$ that minimizes the sum of squares of the differences between Y_i's and their expected values

$$S = \sum \left[Y_i - \mu_i \left(\boldsymbol{\beta} \right) \right]^2.$$

Usually $\widehat{\boldsymbol{\beta}}$ is obtained by differentiating S with respect to each element β_j of $\boldsymbol{\beta}$ and solving the simultaneous equations

$$\frac{\partial S}{\partial \beta_j} = 0, \qquad j = 1, ..., p.$$

Of course it is necessary to check that the solutions correspond to minima (i.e., the matrix of second derivatives is positive definite) and to identify the global minimum from among these solutions and any local minima at the boundary of the parameter space.

Now suppose that the Y_i's have variances σ_i^2 that are not all equal. Then it may be desirable to minimize the weighted sum of squared differences

$$S = \sum w_i \left[Y_i - \mu_i(\boldsymbol{\beta})\right]^2$$

where the weights are $w_i = (\sigma_i^2)^{-1}$. In this way, the observations which are less reliable (that is, the Y_i's with the larger variances) will have less influence on the estimates.

More generally, let $\mathbf{y} = [Y_1, ..., Y_n]^T$ denote a random vector with mean vector $\boldsymbol{\mu} = [\mu_1, ..., \mu_n]^T$ and variance-covariance matrix $\mathbf{V}$. Then the **weighted least squares estimator** is obtained by minimizing

$$S = (\mathbf{y} - \boldsymbol{\mu})^T \mathbf{V}^{-1} (\mathbf{y} - \boldsymbol{\mu}).$$

1.6.4 Comments on estimation.

1. An important distinction between the methods of maximum likelihood and least squares is that the method of least squares can be used without making assumptions about the distributions of the response variables Y_i beyond specifying their expected values and possibly their variance-covariance structure. In contrast, to obtain maximum likelihood estimators we need to specify the joint probability distribution of the Y_i's.
2. For many situations maximum likelihood and least squares estimators are identical.
3. Often numerical methods rather than calculus may be needed to obtain parameter estimates that maximize the likelihood or log-likelihood function or minimize the sum of squares. The following example illustrates this approach.

1.6.5 Example: Tropical cyclones

Table 1.2 shows the number of tropical cyclones in Northeastern Australia for the seasons 1956-7 (season 1) to 1968-9 (season 13), a period of fairly consistent conditions for the definition and tracking of cyclones (Dobson and Stewart, 1974).

Table 1.2 *Numbers of tropical cyclones in 13 successive seasons.*

Season:	1	2	3	4	5	6	7	8	9	10	11	12	13
No. of cyclones	6	5	4	6	6	3	12	7	4	2	6	7	4

Let Y_i denote the number of cyclones in season i, where $i = 1, \ldots, 13$. Suppose the Y_i's are independent random variables with the Poisson distribution

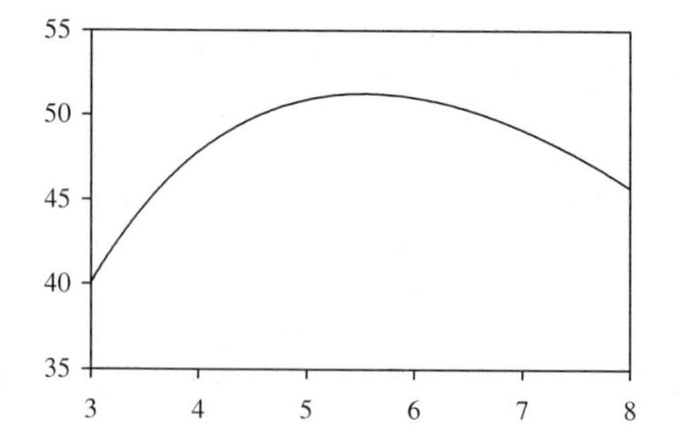

Figure 1.1 *Graph showing the location of the maximum likelihood estimate for the data in Table 1.2 on tropical cyclones.*

with parameter θ. From Example 1.6.2 $\widehat{\theta} = \overline{y} = 72/13 = 5.538$. An alternative approach would be to find numerically the value of θ that maximizes the log-likelihood function. The component of the log-likelihood function due to y_i is

$$l_i = y_i \log \theta - \theta - \log y_i!.$$

The log-likelihood function is the sum of these terms

$$l = \sum_{i=1}^{13} l_i = \sum_{i=1}^{13} \left(y_i \log \theta - \theta - \log y_i! \right).$$

Only the first two terms in the brackets involve θ and so are relevant to the optimization calculation, because the term $\sum_1^{13} \log y_i!$ is a constant. To plot the log-likelihood function (without the constant term) against θ, for various values of θ, calculate $(y_i \log \theta - \theta)$ for each y_i and add the results to obtain $l^* = \sum (y_i \log \theta - \theta)$. Figure 1.1 shows l^* plotted against θ.

Clearly the maximum value is between $\theta = 5$ and $\theta = 6$. This can provide a starting point for an iterative procedure for obtaining $\widehat{\theta}$. The results of a simple bisection calculation are shown in Table 1.3. The function l^* is first calculated for approximations $\theta^{(1)} = 5$ and $\theta^{(2)} = 6$. Then subsequent approximations $\theta^{(k)}$ for $k = 3, 4, ...$ are the average values of the two previous estimates of θ with the largest values of l^*(for example, $\theta^{(6)} = \frac{1}{2}(\theta^{(5)} + \theta^{(3)})$). After 7 steps this process gives $\widehat{\theta} \simeq 5.54$ which is correct to 2 decimal places.

1.7 Exercises

1.1 Let Y_1 and Y_2 be independent random variables with

$Y_1 \sim N(1, 3)$ and $Y_2 \sim N(2, 5)$. If $W_1 = Y_1 + 2Y_2$ and $W_2 = 4Y_1 - Y_2$ what is the joint distribution of W_1 and W_2?

1.2 Let Y_1 and Y_2 be independent random variables with

$Y_1 \sim N(0, 1)$ and $Y_2 \sim N(3, 4)$.

Table 1.3 *Successive approximations to the maximum likelihood estimate of the mean number of cyclones per season.*

k	$\theta^{(k)}$	l^*
1	5	50.878
2	6	51.007
3	5.5	51.242
4	5.75	51.192
5	5.625	51.235
6	5.5625	51.243
7	5.5313	51.24354
8	5.5469	51.24352
9	5.5391	51.24360
10	5.5352	51.24359

(a) What is the distribution of Y_1^2?

(b) If $\mathbf{y} = \begin{bmatrix} Y_1 \\ (Y_2 - 3)/2 \end{bmatrix}$, obtain an expression for $\mathbf{y}^T\mathbf{y}$. What is its distribution?

(c) If $\mathbf{y} = \begin{pmatrix} Y_1 \\ Y_2 \end{pmatrix}$ and its distribution is $\mathbf{y} \sim \mathbf{N}(\boldsymbol{\mu}, \mathbf{V})$, obtain an expression for $\mathbf{y}^T\mathbf{V}^{-1}\mathbf{y}$. What is its distribution?

1.3 Let the joint distribution of Y_1 and Y_2 be $\mathbf{N}(\boldsymbol{\mu}, \mathbf{V})$ with

$$\boldsymbol{\mu} = \begin{pmatrix} 2 \\ 3 \end{pmatrix} \quad \text{and} \quad \mathbf{V} = \begin{pmatrix} 4 & 1 \\ 1 & 9 \end{pmatrix}.$$

(a) Obtain an expression for $(\mathbf{y} - \boldsymbol{\mu})^T\mathbf{V}^{-1}(\mathbf{y} - \boldsymbol{\mu})$. What is its distribution?

(b) Obtain an expression for $\mathbf{y}^T\mathbf{V}^{-1}\mathbf{y}$. What is its distribution?

1.4 Let $Y_1, ..., Y_n$ be independent random variables each with the distribution $N(\mu, \sigma^2)$. Let

$$\overline{Y} = \frac{1}{n}\sum_{i=1}^{n} Y_i \quad \text{and} \quad S^2 = \frac{1}{n-1}\sum_{i=1}^{n}(Y_i - \overline{Y})^2.$$

(a) What is the distribution of $\overline{Y}$?

(b) Show that $S^2 = \dfrac{1}{n-1}\left[\sum_{i=1}^{n}(Y_i - \mu)^2 - n(\overline{Y} - \mu)^2\right]$.

(c) From (b) it follows that $\sum(Y_i - \mu)^2/\sigma^2 = (n-1)S^2/\sigma^2 + \left[(\overline{Y} - \mu)^2 n/\sigma^2\right]$. How does this allow you to deduce that $\overline{Y}$ and S^2 are independent?

(d) What is the distribution of $(n-1)S^2/\sigma^2$?

(e) What is the distribution of $\dfrac{\overline{Y} - \mu}{S/\sqrt{n}}$?

Table 1.4 *Progeny of light brown apple moths.*

Progeny group	Females	Males
1	18	11
2	31	22
3	34	27
4	33	29
5	27	24
6	33	29
7	28	25
8	23	26
9	33	38
10	12	14
11	19	23
12	25	31
13	14	20
14	4	6
15	22	34
16	7	12

1.5 This exercise is a continuation of the example in Section 1.6.2 in which $Y_1, ..., Y_n$ are independent Poisson random variables with the parameter θ.

(a) Show that $E(Y_i) = \theta$ for $i = 1, ..., n$.

(b) Suppose $\theta = e^{\beta}$. Find the maximum likelihood estimator of β.

(c) Minimize $S = \sum \left(Y_i - e^{\beta}\right)^2$ to obtain a least squares estimator of β.

1.6 The data below are the numbers of females and males in the progeny of 16 female light brown apple moths in Muswellbrook, New South Wales, Australia (from Lewis, 1987).

(a) Calculate the proportion of females in each of the 16 groups of progeny.

(b) Let Y_i denote the number of females and n_i the number of progeny in each group ($i = 1, ..., 16$). Suppose the Y_i's are independent random variables each with the binomial distribution

$$f(y_i;\theta) = \binom{n_i}{y_i} \theta^{y_i}(1-\theta)^{n_i - y_i}.$$

Find the maximum likelihood estimator of θ using calculus and evaluate it for these data.

(c) Use a numerical method to estimate $\widehat{\theta}$ and compare the answer with the one from (b).

2

Model Fitting

2.1 Introduction

The model fitting process described in this book involves four steps:

1. Model specification – a model is specified in two parts: an equation linking the response and explanatory variables and the probability distribution of the response variable.
2. Estimation of the parameters of the model.
3. Checking the adequacy of the model – how well it fits or summarizes the data.
4. Inference – calculating confidence intervals and testing hypotheses about the parameters in the model and interpreting the results.

In this chapter these steps are first illustrated using two small examples. Then some general principles are discussed. Finally there are sections about notation and coding of explanatory variables which are needed in subsequent chapters.

2.2 Examples

2.2.1 Chronic medical conditions

Data from the Australian Longitudinal Study on Women's Health (Brown et al., 1996) show that women who live in country areas tend to have fewer consultations with general practitioners (family physicians) than women who live near a wider range of health services. It is not clear whether this is because they are healthier or because structural factors, such as shortage of doctors, higher costs of visits and longer distances to travel, act as barriers to the use of general practitioner (GP) services. Table 2.1 shows the numbers of chronic medical conditions (for example, high blood pressure or arthritis) reported by samples of women living in large country towns (town group) or in more rural areas (country group) in New South Wales, Australia. All the women were aged 70-75 years, had the same socio-economic status and had three or fewer GP visits during 1996. The question of interest is: do women who have similar levels of use of GP services in the two groups have the same need as indicated by their number of chronic medical conditions?

The Poisson distribution provides a plausible way of modelling these data as they are counts and within each group the sample mean and variance are approximately equal. Let Y_{jk} be a random variable representing the number of conditions for the kth woman in the jth group, where $j = 1$ for the town group and $j = 2$ for the country group and $k = 1, \ldots, K_j$ with $K_1 = 26$ and $K_2 = 23$.

Table 2.1 *Numbers of chronic medical conditions of 26 town women and 23 country women with similar use of general practitioner services.*

Town																	
0	1	1	0	2	3	0	1	1	1	1	2	0	1	3	0	1	2
1	3	3	4	1	3	2	0										
n = 26, mean = 1.423, standard deviation = 1.172, variance = 1.374																	
Country																	
2	0	3	0	0	1	1	1	1	0	0	2	2	0	1	2	0	0
1	1	1	0	2													
n = 23, mean = 0.913, standard deviation = 0.900, variance = 0.810																	

Suppose the Y_{jk}'s are all independent and have the Poisson distribution with parameter θ_j representing the expected number of conditions.

The question of interest can be formulated as a test of the null hypothesis $\mathrm{H}_0 : \theta_1 = \theta_2 = \theta$ against the alternative hypothesis $\mathrm{H}_1 : \theta_1 \neq \theta_2$. The model fitting approach to testing H_0 is to fit two models, one assuming H_0 is true, that is

$$\mathrm{E}(Y_{jk}) = \theta; \quad Y_{jk} \sim Poisson(\theta) \tag{2.1}$$

and the other assuming it is not, so that

$$\mathrm{E}(Y_{jk}) = \theta_j; \quad Y_{jk} \sim Poisson(\theta_j), \tag{2.2}$$

where $j = 1$ or 2. Testing H_0 against H_1 involves comparing how well models (2.1) and (2.2) fit the data. If they are about equally good then there is little reason for rejecting H_0. However if model (2.2) is clearly better, then H_0 would be rejected in favor of H_1.

If H_0 is true, then the log-likelihood function of the Y_{jk}'s is

$$l_0 = l(\theta; \mathbf{y}) = \sum_{j=1}^{J} \sum_{k=1}^{K_j} (y_{jk} \log \theta - \theta - \log y_{jk}!), \tag{2.3}$$

where $J = 2$ in this case. The maximum likelihood estimate, which can be obtained as shown in the example in Section 1.6.2, is

$$\widehat{\theta} = \sum \sum y_{jk}/N,$$

where $N = \sum_j K_j$. For these data the estimate is $\widehat{\theta} = 1.184$ and the maximum value of the log-likelihood function, obtained by substituting this value of $\widehat{\theta}$ and the data values y_{jk} into (2.3), is $\widehat{l_0} = -68.3868$.

If H_1 is true, then the log-likelihood function is

$$\begin{aligned} l_1 &= l(\theta_1, \theta_2; \mathbf{y}) = \sum_{k=1}^{K_1} (y_{1k} \log \theta_1 - \theta_1 - \log y_{1k}!) \\ &+ \sum_{k=1}^{K_2} (y_{2k} \log \theta_2 - \theta_2 - \log y_{2k}!). \end{aligned} \tag{2.4}$$

(The subscripts on l_0 and l_1 in (2.3) and (2.4) are used to emphasize the connections with the hypotheses H_0 and H_1, respectively). From (2.4) the maximum likelihood estimates are $\widehat{\theta}_j = \sum_k y_{jk}/K_j$ for $j = 1$ or 2. In this case $\widehat{\theta}_1 = 1.423, \widehat{\theta}_2 = 0.913$ and the maximum value of the log-likelihood function, obtained by substituting these values and the data into (2.4), is $\widehat{l}_1 = -67.0230$.

The maximum value of the log-likelihood function l_1 will always be greater than or equal to that of l_0 because one more parameter has been fitted. To decide whether the difference is statistically significant we need to know the sampling distribution of the log-likelihood function. This is discussed in Chapter 4.

If $Y \sim \text{Poisson}(\theta)$ then $\text{E}(Y) = \text{var}(Y) = \theta$. The estimate $\widehat{\theta}$ of $\text{E}(Y)$ is called the **fitted value** of Y. The difference $Y - \widehat{\theta}$ is called a **residual** (other definitions of residuals are also possible, see Section 2.3.4). Residuals form the basis of many methods for examining the adequacy of a model. A residual is usually standardized by dividing by its standard error. For the Poisson distribution an approximate standardized residual is

$$r = \frac{Y - \widehat{\theta}}{\sqrt{\widehat{\theta}}}.$$

The standardized residuals for models (2.1) and (2.2) are shown in Table 2.2 and Figure 2.1. Examination of individual residuals is useful for assessing certain features of a model such as the appropriateness of the probability distribution used for the responses or the inclusion of specific explanatory variables. For example, the residuals in Table 2.2 and Figure 2.1 exhibit some skewness, as might be expected for the Poisson distribution.

The residuals can also be aggregated to produce summary statistics measuring the overall adequacy of the model. For example, for Poisson data denoted by independent random variables Y_i, provided that the expected values θ_i are not too small, the standardized residuals $r_i = (Y_i - \widehat{\theta}_i)/\sqrt{\widehat{\theta}_i}$ approximately have the standard Normal distribution $N(0,1)$, although they are not usually independent. An intuitive argument is that, approximately, $r_i \sim N(0,1)$ so $r_i^2 \sim \chi^2(1)$ and hence

$$\sum r_i^2 = \sum \frac{(Y_i - \widehat{\theta}_i)^2}{\widehat{\theta}_i} \sim \chi^2(m). \tag{2.5}$$

In fact, it can be shown that for large samples, (2.5) is a good approximation with m equal to the number of observations minus the number of parameters

Table 2.2 *Observed values and standardized residuals for the data on chronic medical conditions (Table 2.1), with estimates obtained from models (2.1) and (2.2).*

Value of Y	Frequency	Standardized residuals from (2.1); $\widehat{\theta} = 1.184$	Standardized residuals from (2.2); $\widehat{\theta}_1 = 1.423$ and $\widehat{\theta}_2 = 0.913$
		Town	
0	6	-1.088	-1.193
1	10	-0.169	-0.355
2	4	0.750	0.484
3	5	1.669	1.322
4	1	2.589	2.160
		Country	
0	9	-1.088	-0.956
1	8	-0.169	0.091
2	5	0.750	1.138
3	1	1.669	2.184

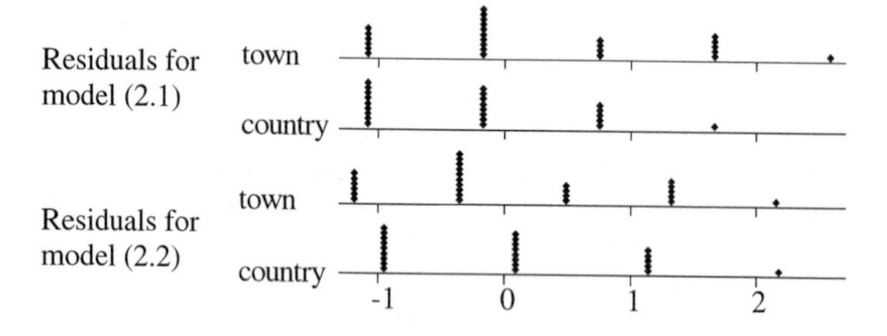

Figure 2.1 *Plots of residuals for models (2.1) and (2.2) for the data in Table 2.2 on chronic medical conditions.*

estimated in order to calculate to fitted values $\widehat{\theta}_i$ (for example, see Agresti, 1990, page 479). Expression (2.5) is, in fact, the usual chi-squared goodness of fit statistic for count data which is often written as

$$X^2 = \sum \frac{(o_i - e_i)^2}{e_i} \sim \chi^2(m)$$

where o_i denotes the observed frequency and e_i denotes the corresponding expected frequency. In this case $o_i = Y_i$, $e_i = \widehat{\theta}_i$ and $\sum r_i^2 = X^2$.

For the data on chronic medical conditions, for model (2.1)

$$\sum r_i^2 = 6 \times (-1.088)^2 + 10 \times (-0.169)^2 + \ldots + 1 \times 1.669^2 = 46.759.$$

This value is consistent with $\sum r_i^2$ being an observation from the central chi-squared distribution with $m = 23 + 26 - 1 = 48$ degrees of freedom. (Recall from Section 1.4.2, that if $X^2 \sim \chi^2(m)$ then $\mathrm{E}(X^2) = m$ and notice that the calculated value $X^2 = \sum r_i^2 = 46.759$ is near the expected value of 48.)

Similarly, for model (2.2)

$$\sum r_i^2 = 6 \times (-1.193)^2 + \ldots + 1 \times 2.184^2 = 43.659$$

which is consistent with the central chi-squared distribution with $m = 49-2 = 47$ degrees of freedom. The difference between the values of $\sum r_i^2$ from models (2.1) and (2.2) is small: $46.759-43.659 = 3.10$. This suggests that model (2.2) with two parameters, may not describe the data much better than the simpler model (2.1). If this is so, then the data provide evidence supporting the null hypothesis $\mathrm{H}_0 : \theta_1 = \theta_2$. More formal testing of the hypothesis is discussed in Chapter 4.

The next example illustrates steps of the model fitting process with continuous data.

2.2.2 Birthweight and gestational age

The data in Table 2.3 are the birthweights (in grams) and estimated gestational ages (in weeks) of 12 male and female babies born in a certain hospital. The mean ages are almost the same for both sexes but the mean birthweight for boys is higher than the mean birthweight for girls. The data are shown in the scatter plot in Figure 2.2. There is a linear trend of birthweight increasing with gestational age and the girls tend to weigh less than the boys of the same gestational age. The question of interest is whether the rate of increase of birthweight with gestational age is the same for boys and girls.

Let Y_{jk} be a random variable representing the birthweight of the kth baby in group j where $j = 1$ for boys and $j = 2$ for girls and $k = 1, \ldots, 12$. Suppose that the Y_{jk}'s are all independent and are Normally distributed with means $\mu_{jk} = \mathrm{E}(Y_{jk})$, which may differ among babies, and variance σ^2 which is the same for all of them.

A fairly general model relating birthweight to gestational age is

$$\mathrm{E}(Y_{jk}) = \mu_{jk} = \alpha_j + \beta_j x_{jk}$$

Table 2.3 *Birthweight and gestational age for boys and girls.*

	Boys		Girls	
	Age	Birthweight	Age	Birthweight
	40	2968	40	3317
	38	2795	36	2729
	40	3163	40	2935
	35	2925	38	2754
	36	2625	42	3210
	37	2847	39	2817
	41	3292	40	3126
	40	3473	37	2539
	37	2628	36	2412
	38	3176	38	2991
	40	3421	39	2875
	38	2975	40	3231
Means	38.33	3024.00	38.75	2911.33

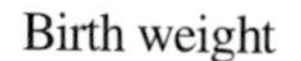

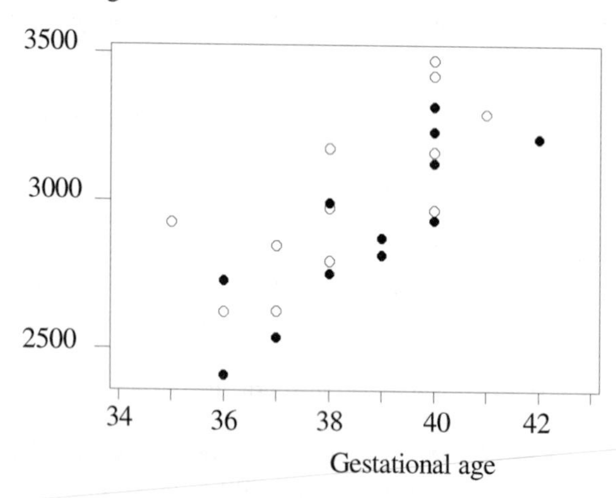

Figure 2.2 *Birthweight plotted against gestational age for boys (open circles) and girls (solid circles); data in Table 2.3.*

where x_{jk} is the gestational age of the kth baby in group j. The intercept parameters α_1 and α_2 are likely to differ because, on average, the boys were heavier than the girls. The slope parameters β_1 and β_2 represent the average increases in birthweight for each additional week of gestational age. The question of interest can be formulated in terms of testing the null hypothesis $\mathrm{H}_0 : \beta_1 = \beta_2 = \beta$ (that is, the growth rates are equal and so the lines are parallel), against the alternative hypothesis $\mathrm{H}_1 : \beta_1 \neq \beta_2$.

We can test H_0 against H_1 by fitting two models

$$\mathrm{E}(Y_{jk}) = \mu_{jk} = \alpha_j + \beta x_{jk}; \quad Y_{jk} \sim N(\mu_{jk}, \sigma^2), \tag{2.6}$$

$$\mathrm{E}(Y_{jk}) = \mu_{jk} = \alpha_j + \beta_j x_{jk}; \quad Y_{jk} \sim N(\mu_{jk}, \sigma^2). \tag{2.7}$$

The probability density function for Y_{jk} is

$$f(y_{jk}; \mu_{jk}) = \frac{1}{\sqrt{2\pi\sigma^2}} \exp[-\frac{1}{2\sigma^2}(y_{jk} - \mu_{jk})^2].$$

We begin by fitting the more general model (2.7). The log-likelihood function is

$$\begin{aligned} l_1(\alpha_1, \alpha_2, \beta_1, \beta_2; \mathbf{y}) &= \sum_{j=1}^{J}\sum_{k=1}^{K}[-\frac{1}{2}\log(2\pi\sigma^2) - \frac{1}{2\sigma^2}(y_{jk} - \mu_{jk})^2] \\ &= -\frac{1}{2}JK\log(2\pi\sigma^2) - \frac{1}{2\sigma^2}\sum_{j=1}^{J}\sum_{k=1}^{K}(y_{jk} - \alpha_j - \beta_j x_{jk})^2 \end{aligned}$$

where $J = 2$ and $K = 12$ in this case. When obtaining maximum likelihood estimates of $\alpha_1, \alpha_2, \beta_1$ and β_2 we treat the parameter σ^2 as a known constant, or **nuisance parameter,** and we do not estimate it.

The maximum likelihood estimates are the solutions of the simultaneous equations

$$\begin{aligned} \frac{\partial l_1}{\partial \alpha_j} &= \frac{1}{\sigma^2}\sum_k (y_{jk} - \alpha_j - \beta_j x_{jk}) = 0, \\ \frac{\partial l_1}{\partial \beta_j} &= \frac{1}{\sigma^2}\sum_k x_{jk}(y_{jk} - \alpha_j - \beta_j x_{jk}) = 0, \end{aligned} \tag{2.8}$$

where $j = 1$ or 2.

An alternative to maximum likelihood estimation is least squares estimation. For model (2.7), this involves minimizing the expression

$$S_1 = \sum_{j=1}^{J}\sum_{k=1}^{K}(y_{jk} - \mu_{jk})^2 = \sum_{j=1}^{J}\sum_{k=1}^{K}(y_{jk} - \alpha_j - \beta_j x_{jk})^2. \tag{2.9}$$

The least squares estimates are the solutions of the equations

$$\begin{aligned}
\frac{\partial S_1}{\partial \alpha_j} &= -2\sum_{k=1}^{K}(y_{jk} - \alpha_j - \beta_j x_{jk}) = 0, \\
\frac{\partial S_1}{\partial \beta_j} &= -2\sum_{k=1}^{K} x_{jk}(y_{jk} - \alpha_j - \beta_j x_{jk}) = 0.
\end{aligned} \tag{2.10}$$

The equations to be solved in (2.8) and (2.10) are the same and so maximizing l_1 is equivalent to minimizing S_1. For the remainder of this example we will use the least squares approach.

The estimating equations (2.10) can be simplified to

$$\begin{aligned}
\sum_{k=1}^{K} y_{jk} - K\alpha_j - \beta_j \sum_{k=1}^{K} x_{jk} &= 0, \\
\sum_{k=1}^{K} x_{jk} y_{jk} - K\alpha_j \sum_{k=1}^{K} x_{jk} - \beta_j \sum_{k=1}^{K} x_{jk}^2 &= 0
\end{aligned}$$

for $j = 1$ or 2. These are called the **normal equations**. The solution is

$$\begin{aligned}
b_j &= \frac{K\sum_k x_{jk}y_{jk} - (\sum_k x_{jk})(\sum_k y_{jk})}{K\sum_k x_{jk}^2 - (\sum_k x_{jk})^2}, \\
a_j &= \overline{y}_j - b_j \overline{x}_j,
\end{aligned}$$

where a_j is the estimate of α_j and b_j is the estimate of β_j, for $j = 1$ or 2. By considering the second derivatives of (2.9) it can be verified that the solution of equations (2.10) does correspond to the minimum of S_1. The numerical value for the minimum value for S_1 for a particular data set can be obtained by substituting the estimates for α_j and β_j and the data values for y_{jk} and x_{jk} into (2.9).

To test $\mathrm{H}_0 : \beta_1 = \beta_2 = \beta$ against the more general alternative hypothesis H_1, the estimation procedure described above for model (2.7) is repeated but with the expression in (2.6) used for μ_{jk}. In this case there are three parameters, α_1, α_2 and β, instead of four to be estimated. The least squares expression to be minimized is

$$S_0 = \sum_{j=1}^{J}\sum_{k=1}^{K}(y_{jk} - \alpha_j - \beta x_{jk})^2. \tag{2.11}$$

From (2.11) the least squares estimates are given by the solution of the simultaneous equations

$$\begin{aligned}
\frac{\partial S_0}{\partial \alpha_j} &= -2\sum_{k=1}^{K}(y_{jk} - \alpha_j - \beta x_{jk}) = 0, \\
\frac{\partial S_0}{\partial \beta} &= -2\sum_{j=1}^{J}\sum_{k=1}^{K} x_{jk}(y_{jk} - \alpha_j - \beta x_{jk}) = 0,
\end{aligned} \tag{2.12}$$

Table 2.4 *Summary of data on birthweight and gestational age in Table 2.3 (summation is over k=1,...,K where K=12).*

	Boys ($j = 1$)	Girls ($j = 2$)
$\sum x$	460	465
$\sum y$	36288	34936
$\sum x^2$	17672	18055
$\sum y^2$	110623496	102575468
$\sum xy$	1395370	1358497

for $j = 1$ and 2. The solution is

$$\begin{aligned} b &= \frac{K \sum_j \sum_k x_{jk} y_{jk} - \sum_j (\sum_k x_{jk} \sum_k y_{jk})}{K \sum_j \sum_k x_{jk}^2 - \sum_j (\sum_k x_{jk})^2}, \\ a_j &= \overline{y}_j - b\overline{x}_j. \end{aligned}$$

These estimates and the minimum value for S_0 can be calculated from the data.

For the example on birthweight and gestational age, the data are summarized in Table 2.4 and the least squares estimates and minimum values for S_0 and S_1 are given in Table 2.5. The fitted values $\widehat{y}_{jk}$ are shown in Table 2.6. For model (2.6), $\widehat{y}_{jk} = a_j + bx_{jk}$ is calculated from the estimates in the top part of Table 2.5. For model (2.7), $\widehat{y}_{jk} = a_j + b_j x_{jk}$ is calculated using estimates in the bottom part of Table 2.5. The residual for each observation is $y_{jk} - \widehat{y}_{jk}$. The standard deviation s of the residuals can be calculated and used to obtain approximate standardized residuals $(y_{jk} - \widehat{y}_{jk})/s$. Figures 2.3 and 2.4 show for models (2.6) and (2.7), respectively: the standardized residuals plotted against the fitted values; the standardized residuals plotted against gestational age; and Normal probability plots. These types of plots are discussed in Section 2.3.4. The Figures show that:

1. Standardized residuals show no systematic patterns in relation to either the fitted values or the explanatory variable, gestational age.
2. Standardized residuals are approximately Normally distributed (as the points are near the solid lines in the bottom graphs).
3. Very little difference exists between the two models.

The apparent lack of difference between the models can be examined by testing the null hypothesis H_0 (corresponding to model (2.6)) against the alternative hypothesis H_1 (corresponding to model (2.7)). If H_0 is correct, then the minimum values $\widehat{S}_1$ and $\widehat{S}_0$ should be nearly equal. If the data support this hypothesis, we would feel justified in using the simpler model (2.6) to describe the data. On the other hand, if the more general hypothesis H_1 is true then $\widehat{S}_0$ should be much larger than $\widehat{S}_1$ and model (2.7) would be preferable.

To assess the relative magnitude of the values $\widehat{S}_1$ and $\widehat{S}_0$ we need to use the

Table 2.5 *Analysis of data on birthweight and gestational age in Table 2.3.*

Model	Slopes	Intercepts	Minimum sum of squares
(2.6)	$b = 120.894$	$a_1 = -1610.283$	$\widehat{S}_0 = 658770.8$
		$a_2 = -1773.322$	
(2.7)	$b_1 = 111.983$	$a_1 = -1268.672$	$\widehat{S}_1 = 652424.5$
	$b_2 = 130.400$	$a_2 = -2141.667$	

Table 2.6 *Observed values and fitted values under model (2.6) and model (2.7) for data in Table 2.3.*

Sex	Gestational age	Birthweight	Fitted value under (2.6)	Fitted value under (2.7)
Boys	40	2968	3225.5	3210.6
	38	2795	2983.7	2986.7
	40	3163	3225.5	3210.6
	35	2925	2621.0	2650.7
	36	2625	2741.9	2762.7
	37	2847	2862.8	2874.7
	41	3292	3346.4	3322.6
	40	3473	3225.5	3210.6
	37	2628	2862.8	2874.7
	38	3176	2983.7	2986.7
	40	3421	3225.5	3210.6
	38	2975	2983.7	2986.7
Girls	40	3317	3062.5	3074.3
	36	2729	2578.9	2552.7
	40	2935	3062.5	3074.3
	38	2754	2820.7	2813.5
	42	3210	3304.2	3335.1
	39	2817	2941.6	2943.9
	40	3126	3062.5	3074.3
	37	2539	2699.8	2683.1
	36	2412	2578.9	2552.7
	38	2991	2820.7	2813.5
	39	2875	2941.6	2943.9
	40	3231	3062.5	3074.3

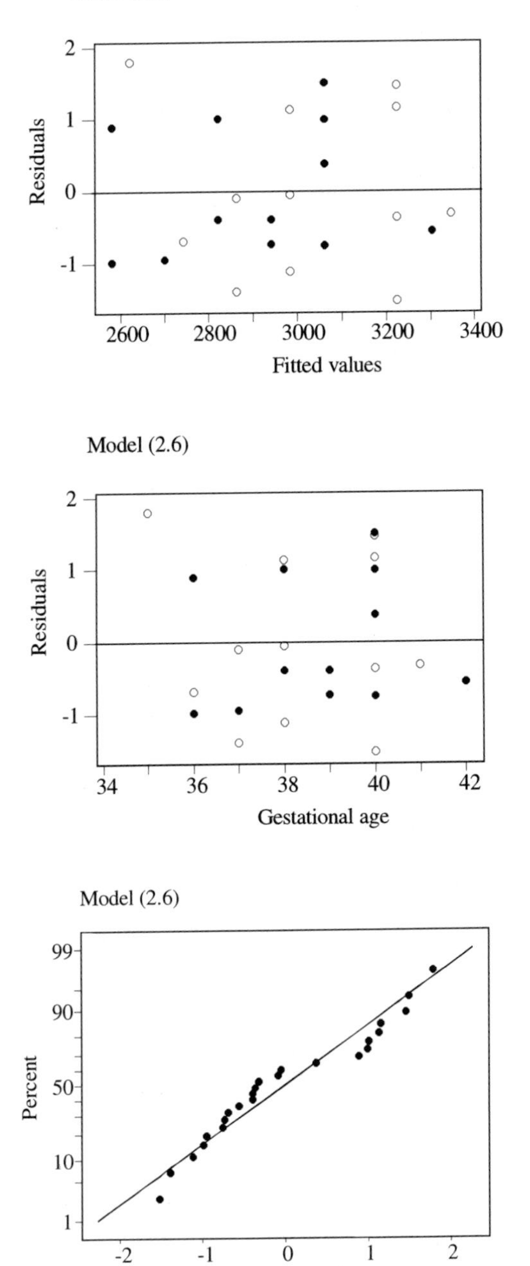

Figure 2.3 *Plots of standardized residuals for Model (2.6) for the data on birthweight and gestational age (Table 2.3); for the top and middle plots, open circles correspond to data from boys and solid circles correspond to data from girls.*

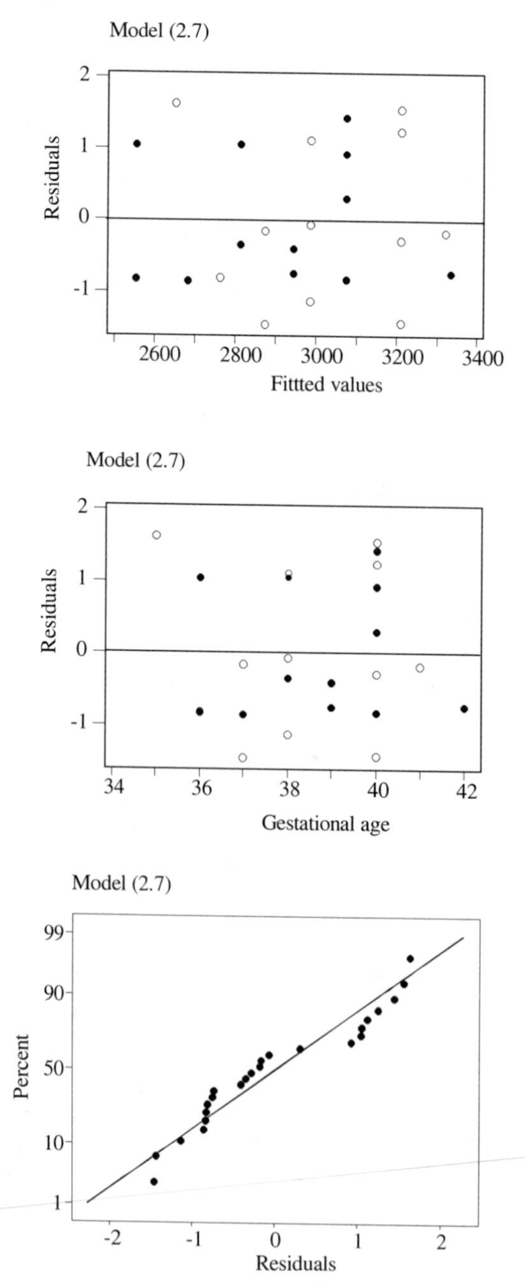

Figure 2.4 *Plots of standardized residuals for Model (2.7) for the data on birthweight and gestational age (Table 2.3); for the top and middle plots, open circles correspond to data from boys and solid circles correspond to data from girls.*

sampling distributions of the corresponding random variables

$$\widehat{S}_1 = \sum_{j=1}^{J}\sum_{k=1}^{K}(Y_{jk} - a_j - b_j x_{jk})^2$$

and

$$\widehat{S}_0 = \sum_{j=1}^{J}\sum_{k=1}^{K}(Y_{jk} - a_j - b x_{jk})^2.$$

It can be shown (see Exercise 2.3) that

$$\begin{aligned}\widehat{S}_1 &= \sum_{j=1}^{J}\sum_{k=1}^{K}[Y_{jk} - (\alpha_j + \beta_j x_{jk})]^2 - K\sum_{j=1}^{J}(\overline{Y}_j - \alpha_j - \beta_j \overline{x}_j)^2 \\ &\quad - \sum_{j=1}^{J}(b_j - \beta_j)^2(\sum_{k=1}^{K} x_{jk}^2 - K\overline{x}_j^2)\end{aligned}$$

and that the random variables Y_{jk}, $\overline{Y}_j$ and b_j are all independent and have the following distributions:

$$\begin{aligned}Y_{jk} &\sim N(\alpha_j + \beta_j x_{jk}, \sigma^2), \\ \overline{Y}_j &\sim N(\alpha_j + \beta_j \overline{x}_j, \sigma^2/K), \\ b_j &\sim N(\beta_j, \sigma^2/(\sum_{k=1}^{K} x_{jk}^2 - K\overline{x}_j^2)).\end{aligned}$$

Therefore $\widehat{S}_1/\sigma^2$ is a linear combination of sums of squares of random variables with Normal distributions. In general, there are JK random variables $(Y_{jk} - \alpha_j - \beta_j x_{jk})^2/\sigma^2$, J random variables $(\overline{Y}_j - \alpha_j - \beta_j \overline{x}_j)^2 K/\sigma^2$ and J random variables $(b_j - \beta_j)^2(\sum_k x_{jk}^2 - K\overline{x}_j^2)/\sigma^2$. They are all independent and each has the $\chi^2(1)$ distribution. From the properties of the chi-squared distribution in Section 1.5, it follows that $\widehat{S}_1/\sigma^2 \sim \chi^2(JK - 2J)$. Similarly, if H_0 is correct then $\widehat{S}_0/\sigma^2 \sim \chi^2[JK - (J+1)]$. In this example $J = 2$ so $\widehat{S}_1/\sigma^2 \sim \chi^2(2K - 4)$ and $\widehat{S}_0/\sigma^2 \sim \chi^2(2K - 3)$. In each case the value for the degrees of freedom is the number of observations minus the number of parameters estimated.

If β_1 and β_2 are not equal (corresponding to H_1), then $\widehat{S}_0/\sigma^2$ will have a non-central chi-squared distribution with $JK - (J+1)$ degrees of freedom. On the other hand, provided that model (2.7) describes the data well, $\widehat{S}_1/\sigma^2$ will have a central chi-squared distribution with $JK - 2J$ degrees of freedom.

The statistic $\widehat{S}_0 - \widehat{S}_1$ represents the improvement in fit of (2.7) compared to (2.6). If H_0 is correct, then

$$\frac{1}{\sigma^2}(\widehat{S}_0 - \widehat{S}_1) \sim \chi^2(J-1).$$

If H_0 is not correct then $(\widehat{S}_0 - \widehat{S}_1)/\sigma^2$ has a non-central chi-squared distribu-

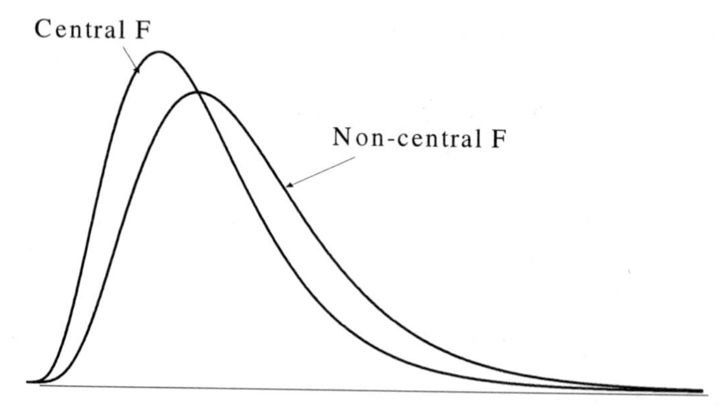

Figure 2.5 *Central and non-central F distributions.*

tion. However, as σ^2 is unknown, we cannot compare $(\widehat{S}_0 - \widehat{S}_1)/\sigma^2$ directly with the $\chi^2(J-1)$ distribution. Instead we eliminate σ^2 by using the ratio of $(\widehat{S}_0 - \widehat{S}_1)/\sigma^2$ and the random variable $\widehat{S}_1/\sigma^2$ with a central chi-squared distribution, each divided by the relevant degrees of freedom,

$$F = \frac{(\widehat{S}_0 - \widehat{S}_1)/\sigma^2}{(J-1)} / \frac{\widehat{S}_1/\sigma^2}{(JK-2J)} = \frac{(\widehat{S}_0 - \widehat{S}_1)/(J-1)}{\widehat{S}_1/(JK-2J)}.$$

If H_0 is correct, from Section 1.4.4, F has the central distribution $F(J-1, JK-2J)$. If H_0 is not correct, F has a non-central F-distribution and the calculated value of F will be larger than expected from the central F-distribution (see Figure 2.5).

For the example on birthweight and gestational age, the value of F is

$$\frac{(658770.8 - 652424.5)/1}{652424.5/20} = 0.19$$

This value is certainly not statistically significant when compared with the $F(1,20)$ distribution. Thus the data do not provide evidence against the hypothesis $\mathrm{H}_0 : \beta_0 = \beta_1$, and on the grounds of simplicity model (2.6), which specifies the same slopes but different intercepts, is preferable.

These two examples illustrate the main ideas and methods of statistical modelling which are now discussed more generally.

2.3 Some principles of statistical modelling

2.3.1 Exploratory data analysis

Any analysis of data should begin with a consideration of each variable separately, both to check on data quality (for example, are the values plausible?) and to help with model formulation.

1. What is the scale of measurement? Is it continuous or categorical? If it

is categorical how many categories does it have and are they nominal or ordinal?

2. What is the shape of the distribution? This can be examined using frequency tables, dot plots, histograms and other graphical methods.

3. How is it associated with other variables? Cross tabulations for categorical variables, scatter plots for continuous variables, side-by-side box plots for continuous scale measurements grouped according to the factor levels of a categorical variable, and other such summaries can help to identify patterns of association. For example, do the points on a scatter plot suggest linear or non-linear relationships? Do the group means increase or decrease consistently with an ordinal variable defining the groups?

2.3.2 Model formulation

The models described in this book involve a single response variable Y and usually several explanatory variables. Knowledge of the context in which the data were obtained, including the substantive questions of interest, theoretical relationships among the variables, the study design and results of the exploratory data analysis can all be used to help formulate a model. The model has two components:

1. Probability distribution of Y, for example, $Y \sim N(\mu, \sigma^2)$.

2. Equation linking the expected value of Y with a linear combination of the explanatory variables, for example, $\mathrm{E}(Y) = \alpha + \beta x$ or $\ln[\mathrm{E}(Y)] = \beta_0 + \beta_1 \sin(\alpha x)$.

For generalized linear models the probability distributions all belong to the exponential family of distributions, which includes the Normal, binomial, Poisson and many other distributions. This family of distributions is discussed in Chapter 3. The equation in the second part of the model has the general form

$$g[\mathrm{E}(Y)] = \beta_0 + \beta_1 x_1 + \ldots + \beta_m x_m$$

where the part $\beta_0 + \beta_1 x_1 + \ldots + \beta_m x_m$ is called the **linear component**. Notation for the linear component is discussed in Section 2.4.

2.3.3 Parameter estimation

The most commonly used estimation methods are maximum likelihood and least squares. These are described in Section 1.6. In this book numerical and graphical methods are used, where appropriate, to complement calculus and algebraic methods of optimization.

2.3.4 Residuals and model checking

Firstly, consider residuals for a model involving the Normal distribution. Suppose that the response variable Y_i is modelled by

$$\mathrm{E}(Y_i) = \mu_i; \quad Y_i \sim N(\mu_i, \sigma^2).$$

The fitted values are the estimates $\widehat{\mu}_i$. Residuals can be defined as $y_i - \widehat{\mu}_i$ and the approximate standardized residuals as

$$r_i = (y_i - \widehat{\mu}_i)/\widehat{\sigma},$$

where $\widehat{\sigma}$ is an estimate of the unknown parameter σ. These standardized residuals are slightly correlated because they all depend on the estimates $\widehat{\mu}_i$ and $\widehat{\sigma}$ that were calculated from the observations. Also they are not exactly Normally distributed because σ has been estimated by $\widehat{\sigma}$. Nevertheless, they are approximately Normally distributed and the adequacy of the approximation can be checked using appropriate graphical methods (see below).

The parameters μ_i are functions of the explanatory variables. If the model is a good description of the relationship between the response and the explanatory variables, this should be well 'captured' or 'explained' by the $\widehat{\mu}_i$'s. Therefore there should be little remaining information in the residuals $y_i - \widehat{\mu}_i$. This too can be checked graphically (see below). Additionally, the sum of squared residuals $\sum(y_i - \widehat{\mu}_i)^2$ provides an overall statistic for assessing the adequacy of the model; in fact, it is the component of the log-likelihood function or least squares expression which is optimized in the estimation process.

Secondly, consider residuals from a Poisson model. Recall the model for chronic medical conditions

$$\mathrm{E}(Y_i) = \theta_i; \quad Y_i \sim Poisson(\theta_i).$$

In this case approximate standardized residuals are of the form

$$r_i = \frac{y_i - \widehat{\theta}_i}{\sqrt{\widehat{\theta}_i}}.$$

These can be regarded as signed square roots of contributions to the Pearson goodness-of-fit statistic

$$\sum_i \frac{(o_i - e_i)^2}{e_i},$$

where o_i is the observed value y_i and e_i is the fitted value $\widehat{\theta}_i$ 'expected' from the model.

For other distributions a variety of definitions of standardized residuals are used. Some of these are transformations of the terms $(y_i - \widehat{\mu}_i)$ designed to improve their Normality or independence (for example, see Chapter 9 of Neter et al., 1996). Others are based on signed square roots of contributions to statistics, such as the log-likelihood function or the sum of squares, which are used as overall measures of the adequacy of the model (for example, see

Cox and Snell, 1968; Prigibon, 1981; and Pierce and Shafer, 1986). Many of these residuals are discussed in more detail in McCullagh and Nelder (1989) or Krzanowski (1998).

Residuals are important tools for checking the assumptions made in formulating a model. This is because they should usually be independent and have a distribution which is approximately Normal with a mean of zero and constant variance. They should also be unrelated to the explanatory variables. Therefore, the standardized residuals can be compared to the Normal distribution to assess the adequacy of the distributional assumptions and to identify any unusual values. This can be done by inspecting their frequency distribution and looking for values beyond the likely range; for example, no more than 5% should be less than -1.96 or greater than $+1.96$ and no more than 1% should be beyond ± 2.58.

A more sensitive method for assessing Normality, however, is to use a **Normal probability plot**. This involves plotting the residuals against their expected values, defined according to their rank order, if they were Normally distributed. These values are called the **Normal order statistics** and they depend on the number of observations. Normal probability plots are available in all good statistical software (and analogous probability plots for other distributions are also commonly available). In the plot the points should lie on or near a straight line representing Normality and systematic deviations or outlying observations indicate a departure from this distribution.

The standardized residuals should also be plotted against each of the explanatory variables that are included in the model. If the model adequately describes the effect of the variable, there should be no apparent pattern in the plot. If it is inadequate, the points may display curvature or some other systematic pattern which would suggest that additional or alternative terms may need to be included in the model. The residuals should also be plotted against other potential explanatory variables that are not in the model. If there is any systematic pattern, this suggests that additional variables should be included. Several different residual plots for detecting non-linearity in generalized linear models have been compared by Cai and Tsai (1999).

In addition, the standardized residuals should be plotted against the fitted values $\widehat{y}_i$, especially to detect changes in variance. For example, an increase in the spread of the residuals towards the end of the range of fitted values would indicate a departure from the assumption of constant variance (sometimes termed **homoscedasticity**).

Finally, a sequence plot of the residuals should be made using the order in which the values y_i were measured. This might be in time order, spatial order or any other sequential effect that might cause lack of independence among the observations. If the residuals are independent the points should fluctuate randomly without any systematic pattern, such as alternating up and down or steadily increasing or decreasing. If there is evidence of associations among the residuals, this can be checked by calculating serial correlation coefficients among them. If the residuals are correlated, special modelling methods are needed – these are outlined in Chapter 11.

2.3.5 *Inference and interpretation*

It is sometimes useful to think of scientific data as measurements composed of a message, or **signal,** that is distorted by **noise.** For instance, in the example about birthweight the 'signal' is the usual growth rate of babies and the 'noise' comes from all the genetic and environmental factors that lead to individual variation. A goal of statistical modelling is to extract as much information as possible about the signal. In practice, this has to be balanced against other criteria such as simplicity. The Oxford Dictionary describes the **law of parsimony** (otherwise known as **Occam's Razor**) as the principle that no more causes should be assumed than will account for the effect. Accordingly a simpler or more parsimonious model that describes the data adequately is preferable to a more complicated one which leaves little of the variability 'unexplained'. To determine a parsimonious model consistent with the data, we test hypotheses about the parameters.

Hypothesis testing is performed in the context of model fitting by defining a series of nested models corresponding to different hypotheses. Then the question about whether the data support a particular hypothesis can be formulated in terms of the adequacy of fit of the corresponding model relative to other more complicated models. This logic is illustrated in the examples earlier in this chapter. Chapter 5 provides a more detailed explanation of the concepts and methods used, including the sampling distributions for the statistics used to describe 'goodness of fit'.

While hypothesis testing is useful for identifying a good model, it is much less useful for interpreting it. Wherever possible, the parameters in a model should have some natural interpretation; for example, the rate of growth of babies, the relative risk of acquiring a disease or the mean difference in profit from two marketing strategies. The estimated magnitude of the parameter and the reliability of the estimate as indicated by its standard error or a confidence interval are far more informative than significance levels or p-values. They make it possible to answer questions such as: is the effect estimated with sufficient precision to be useful, or is the effect large enough to be of practical, social or biological significance?

2.3.6 *Further reading*

An excellent discussion of the principles of statistical modelling is in the introductory part of Cox and Snell (1981). The importance of adopting a systematic approach is stressed by Kleinbaum et al. (1998). The various steps of model choice, criticism and validation are outlined by Krzanowski (1998). The use of residuals is described in Neter et al. (1996), Draper and Smith (1998), Belsley et al. (1980) and Cook and Weisberg (1999).

2.4 Notation and coding for explanatory variables

For the models in this book the equation linking each response variable Y and a set of explanatory variables $x_1, x_2, \ldots x_m$ has the form

$$g[\mathrm{E}(Y)] = \beta_0 + \beta_1 x_1 + \ldots + \beta_m x_m.$$

For responses $Y_1, ..., Y_N$, this can be written in matrix notation as

$$g[\mathrm{E}(\mathbf{y})] = \mathbf{X}\boldsymbol{\beta} \tag{2.13}$$

where

$$\mathbf{y} = \begin{bmatrix} Y_1 \\ . \\ . \\ . \\ Y_N \end{bmatrix} \quad \text{is a vector of responses,}$$

$$g[E(\mathbf{y})] = \begin{bmatrix} g[\mathrm{E}(Y_1)] \\ . \\ . \\ . \\ g[\mathrm{E}(Y_N)] \end{bmatrix}$$

denotes a vector of functions of the terms $\mathrm{E}(Y_i)$ (with the same g for every element),

$$\boldsymbol{\beta} = \begin{bmatrix} \beta_1 \\ . \\ . \\ . \\ \beta_p \end{bmatrix} \quad \text{is a vector of parameters,}$$

and $\mathbf{X}$ is a matrix whose elements are constants representing levels of categorical explanatory variables or measured values of continuous explanatory variables.

For a continuous explanatory variable x (such as gestational age in the example on birthweight) the model contains a term βx where the parameter β represents the change in the response corresponding to a change of one unit in x.

For categorical explanatory variables there are parameters for the different levels of a factor. The corresponding elements of $\mathbf{X}$ are chosen to exclude or include the appropriate parameters for each observation; they are called **dummy variables**. If they are only zeros and ones, the term **indictor variable** is used.

If there are p parameters in the model and N observations, then $\mathbf{y}$ is a $N \times 1$ random vector, $\boldsymbol{\beta}$ is a $p \times 1$ vector of parameters and $\mathbf{X}$ is an $N \times p$ matrix of known constants. $\mathbf{X}$ is often called the **design matrix** and $\mathbf{X}\boldsymbol{\beta}$ is the **linear component** of the model. Various ways of defining the elements of $\mathbf{X}$ are illustrated in the following examples.

2.4.1 Example: Means for two groups

For the data on chronic medical conditions the equation in the model

$$\mathrm{E}(Y_{jk}) = \theta_j; \quad Y_{jk} \sim Poisson(\theta_j), j = 1, 2$$

can be written in the form of (2.13) with g as the identity function, (i.e., $g(\theta_j) = \theta_j$),

$$\mathbf{y} = \begin{bmatrix} Y_{1,1} \\ Y_{1,2} \\ \vdots \\ Y_{1,26} \\ Y_{2,1} \\ \vdots \\ Y_{2,23} \end{bmatrix}, \quad \boldsymbol{\beta} = \begin{bmatrix} \theta_1 \\ \theta_2 \end{bmatrix} \quad \text{and} \quad \mathbf{X} = \begin{bmatrix} 1 & 0 \\ 1 & 0 \\ \vdots & \vdots \\ 1 & 0 \\ 0 & 1 \\ \vdots & \vdots \\ 0 & 1 \end{bmatrix}$$

The top part of $\mathbf{X}$ picks out the terms θ_1 corresponding to $\mathrm{E}(Y_{1k})$ and the bottom part picks out θ_2 for $\mathrm{E}(Y_{2k})$. With this model the group means θ_1 and θ_2 can be estimated and compared.

2.4.2 Example: Simple linear regression for two groups

The more general model for the data on birthweight and gestational age is

$$\mathrm{E}(Y_{jk}) = \mu_{jk} = \alpha_j + \beta_j x_{jk}; \quad Y_{jk} \sim N(\mu_{jk}, \sigma^2).$$

This can be written in the form of (2.13) if g is the identity function,

$$\mathbf{y} = \begin{bmatrix} Y_{11} \\ Y_{12} \\ \vdots \\ Y_{1K} \\ Y_{21} \\ \vdots \\ Y_{2K} \end{bmatrix}, \quad \boldsymbol{\beta} = \begin{bmatrix} \alpha_1 \\ \alpha_2 \\ \beta_1 \\ \beta_2 \end{bmatrix} \quad \text{and} \quad \mathbf{X} = \begin{bmatrix} 1 & 0 & x_{11} & 0 \\ 1 & 0 & x_{12} & 0 \\ \vdots & \vdots & \vdots & \vdots \\ 1 & 0 & x_{1K} & 0 \\ 0 & 1 & 0 & x_{21} \\ \vdots & \vdots & \vdots & \vdots \\ 0 & 1 & 0 & x_{2K} \end{bmatrix}$$

2.4.3 Example: Alternative formulations for comparing the means of two groups

There are several alternative ways of formulating the linear components for comparing means of two groups: $Y_{11}, ..., Y_{1K_1}$ and $Y_{21}, ..., Y_{2K_2}$.

(a) $\mathrm{E}(Y_{1k}) = \beta_1$, and $\mathrm{E}(Y_{2k}) = \beta_2$.

This is the version used in Example 2.4.1 above. In this case $\boldsymbol{\beta} = \begin{bmatrix} \beta_1 \\ \beta_2 \end{bmatrix}$

and the rows of $\mathbf{X}$ are as follows

$$\begin{array}{lcl} Group\ 1 & : & \left[\begin{array}{cc} 1 & 0 \end{array}\right] \\ Group\ 2 & : & \left[\begin{array}{cc} 0 & 1 \end{array}\right]. \end{array}$$

(b) $\mathrm{E}(Y_{1k}) = \mu + \alpha_1$, and $\mathrm{E}(Y_{2k}) = \mu + \alpha_2$.

In this version μ represents the overall mean and α_1 and α_2 are the group differences from μ. In this case $\boldsymbol{\beta} = \left[\begin{array}{c} \mu \\ \alpha_1 \\ \alpha_2 \end{array}\right]$ and the rows of $\mathbf{X}$ are

$$\begin{array}{lcl} Group\ 1 & : & \left[\begin{array}{ccc} 1 & 1 & 0 \end{array}\right] \\ Group\ 2 & : & \left[\begin{array}{ccc} 1 & 0 & 1 \end{array}\right]. \end{array}$$

This formulation, however, has too many parameters as only two parameters can be estimated from the two sets of observations. Therefore some modification or constraint is needed.

(c) $\mathrm{E}(Y_{1k}) = \mu$ and $\mathrm{E}(Y_{2k}) = \mu + \alpha$.

Here Group 1 is treated as the reference group and α represents the additional effect of Group 2. For this version $\boldsymbol{\beta} = \left[\begin{array}{c} \mu \\ \alpha \end{array}\right]$ and the rows of $\mathbf{X}$ are

$$\begin{array}{lcl} Group\ 1 & : & \left[\begin{array}{cc} 1 & 0 \end{array}\right] \\ Group\ 2 & : & \left[\begin{array}{cc} 1 & 1 \end{array}\right]. \end{array}$$

This is an example of **corner point parameterization** in which group effects are defined as differences from a reference category called the 'corner point'.

(d) $\mathrm{E}(Y_{1k}) = \mu + \alpha$, and $\mathrm{E}(Y_{2k}) = \mu - \alpha$.

This version treats the two groups symmetrically; μ is the overall average effect and α represents the group differences. This is an example of a **sum-to-zero constraint** because

$$[\mathrm{E}(Y_{1k}) - \mu] + [\mathrm{E}(Y_{2k}) - \mu] = \alpha + (-\alpha) = 0.$$

In this case $\boldsymbol{\beta} = \left[\begin{array}{c} \mu \\ \alpha \end{array}\right]$ and the rows of $\mathbf{X}$ are

$$\begin{array}{lcl} Group\ 1 & : & \left[\begin{array}{cc} 1 & 1 \end{array}\right] \\ Group\ 2 & : & \left[\begin{array}{cc} 1 & -1 \end{array}\right]. \end{array}$$

2.4.4 Example: Ordinal explanatory variables

Let Y_{jk} denote a continuous measurement of quality of life. Data are collected for three groups of patients with mild, moderate or severe disease. The groups can be described by levels of an ordinal variable. This can be specified by defining the model using

$$\begin{aligned} \mathrm{E}(Y_{1k}) &= \mu \\ \mathrm{E}(Y_{2k}) &= \mu + \alpha_1 \\ \mathrm{E}(Y_{3k}) &= \mu + \alpha_1 + \alpha_2 \end{aligned}$$

and hence $\boldsymbol{\beta} = \begin{bmatrix} \mu \\ \alpha_1 \\ \alpha_2 \end{bmatrix}$ and the rows of $\mathbf{X}$ are

$$\begin{aligned} Group\ 1 &: \begin{bmatrix} 1 & 0 & 0 \end{bmatrix} \\ Group\ 2 &: \begin{bmatrix} 1 & 1 & 0 \end{bmatrix} \\ Group\ 3 &: \begin{bmatrix} 1 & 1 & 1 \end{bmatrix}. \end{aligned}$$

Thus α_1 represents the effect of Group 2 relative to Group 1 and α_2 represents the effect of Group 3 relative to Group 2.

2.5 Exercises

2.1 Genetically similar seeds are randomly assigned to be raised in either a nutritionally enriched environment (treatment group) or standard conditions (control group) using a completely randomized experimental design. After a predetermined time all plants are harvested, dried and weighed. The results, expressed in grams, for 20 plants in each group are shown in Table 2.7.

Table 2.7 *Dried weight of plants grown under two conditions.*

Treatment group		Control group	
4.81	5.36	4.17	4.66
4.17	3.48	3.05	5.58
4.41	4.69	5.18	3.66
3.59	4.44	4.01	4.50
5.87	4.89	6.11	3.90
3.83	4.71	4.10	4.61
6.03	5.48	5.17	5.62
4.98	4.32	3.57	4.53
4.90	5.15	5.33	6.05
5.75	6.34	5.59	5.14

We want to test whether there is any difference in yield between the two groups. Let Y_{jk} denote the kth observation in the jth group where $j = 1$ for the treatment group, $j = 2$ for the control group and $k = 1, ..., 20$ for both groups. Assume that the Y_{jk}'s are independent random variables with $Y_{jk} \sim N(\mu_j, \sigma^2)$. The null hypothesis $\mathrm{H}_0 : \mu_1 = \mu_2 = \mu$, that there is no difference, is to be compared to the alternative hypothesis $\mathrm{H}_1 : \mu_1 \neq \mu_2$.

(a) Conduct an exploratory analysis of the data looking at the distributions for each group (e.g., using dot plots, stem and leaf plots or Normal probability plots) and calculating summary statistics (e.g., means, medians, standard derivations, maxima and minima). What can you infer from these investigations?

(b) Perform an unpaired t-test on these data and calculate a 95% confidence interval for the difference between the group means. Interpret these results.

(c) The following models can be used to test the null hypothesis H_0 against the alternative hypothesis H_1, where

$$\begin{aligned} \mathrm{H}_0 &: \mathrm{E}(Y_{jk}) = \mu; \quad Y_{jk} \sim N(\mu, \sigma^2), \\ \mathrm{H}_1 &: \mathrm{E}(Y_{jk}) = \mu_j; \quad Y_{jk} \sim N(\mu_j, \sigma^2), \end{aligned}$$

for $j = 1, 2$ and $k = 1, ..., 20$. Find the maximum likelihood and least squares estimates of the parameters μ, μ_1 and μ_2, assuming σ^2 is a known constant.

(d) Show that the minimum values of the least squares criteria are:

$$\text{for } \mathrm{H}_0, \ \widehat{S}_0 = \sum\sum(Y_{jk} - \overline{Y})^2 \text{ where } \overline{Y} = \sum_{k=1}^{K}\sum_{k=1}^{K} Y_{jk}/40,$$

$$\text{for } \mathrm{H}_1, \ \widehat{S}_1 = \sum\sum(Y_{jk} - \overline{Y}_j)^2 \text{ where } \overline{Y}_j = \sum_{k=1}^{K} Y_{jk}/20$$

for $j = 1, 2$.

(e) Using the results of Exercise 1.4 show that

$$\frac{1}{\sigma^2}\widehat{S}_1 = \frac{1}{\sigma^2}\sum_{k=1}^{20}\sum_{k=1}^{20}(Y_{jk} - \mu_j)^2 - \frac{20}{\sigma^2}\sum_{k=1}^{2}(\overline{Y}_j - \mu_j)^2$$

and deduce that if H_1 is true

$$\frac{1}{\sigma^2}\widehat{S}_1 \sim \chi^2(38).$$

Similarly show that

$$\frac{1}{\sigma^2}\widehat{S}_0 = \frac{1}{\sigma^2}\sum_{j=1}^{2}\sum_{k=1}^{20}(Y_{jk} - \mu)^2 - \frac{40}{\sigma^2}\sum_{j=1}^{2}(\overline{Y} - \mu)^2$$

and if H_0 is true then

$$\frac{1}{\sigma^2}\widehat{S}_0 \sim \chi^2(39).$$

(f) Use an argument similar to the one in Example 2.2.2 and the results from (e) to deduce that the statistic

$$F = \frac{\widehat{S}_0 - \widehat{S}_1}{\widehat{S}_1/38}$$

has the central F-distribution $F(1, 38)$ if H_0 is true and a non-central distribution if H_0 is not true.

(g) Calculate the F-statistic from (f) and use it to test H_0 against H_1. What do you conclude?

(h) Compare the value of F-statistic from (g) with the t-statistic from (b), recalling the relationship between the t-distribution and the F-distribution (see Section 1.4.4) Also compare the conclusions from (b) and (g).

(i) Calculate residuals from the model for H_0 and use them to explore the distributional assumptions.

2.2 The weights, in kilograms, of twenty men before and after participation in a 'waist loss' program are shown in Table 2.8. (Egger et al., 1999) We want to know if, on average, they retain a weight loss twelve months after the program.

Table 2.8 *Weights of twenty men before and after participation in a 'waist loss' program.*

Man	Before	After	Man	Before	After
1	100.8	97.0	11	105.0	105.0
2	102.0	107.5	12	85.0	82.4
3	105.9	97.0	13	107.2	98.2
4	108.0	108.0	14	80.0	83.6
5	92.0	84.0	15	115.1	115.0
6	116.7	111.5	16	103.5	103.0
7	110.2	102.5	17	82.0	80.0
8	135.0	127.5	18	101.5	101.5
9	123.5	118.5	19	103.5	102.6
10	95.0	94.2	20	93.0	93.0

Let Y_{jk} denote the weight of the kth man at the jth time where $j = 1$ before the program and $j = 2$ twelve months later. Assume the Y_{jk}'s are independent random variables with $Y_{jk} \sim N(\mu_j, \sigma^2)$ for $j = 1, 2$ and $k = 1, ..., 20$.

(a) Use an unpaired t-test to test the hypothesis

$$\mathrm{H}_0 : \mu_1 = \mu_2 \quad \text{versus} \quad \mathrm{H}_1 : \mu_1 \neq \mu_2.$$

(b) Let $D_k = Y_{1k} - Y_{2k}$, for $k = 1, ..., 20$. Formulate models for testing H_0 against H_1 using the D_k's. Using analogous methods to Exercise 2.1 above, assuming σ^2 is a known constant, test H_0 against H_1.

(c) The analysis in (b) is a paired t-test which uses the natural relationship between weights of the *same* person before and after the program. Are the conclusions the same from (a) and (b)?

(d) List the assumptions made for (a) and (b). Which analysis is more appropriate for these data?

2.3 For model (2.7) for the data on birthweight and gestational age, using methods similar to those for Exercise 1.4, show

$$\begin{aligned}
\widehat{S}_1 &= \sum_{j=1}^{J}\sum_{k=1}^{K}(Y_{jk} - a_j - b_j x_{jk})^2 \\
&= \sum_{j=1}^{J}\sum_{k=1}^{K}[(Y_{jk} - (\alpha_j + \beta_j x_{jk})]^2 - K\sum_{j=1}^{J}(\overline{Y}_j - \alpha_j - \beta_j \overline{x}_j)^2 \\
&\quad - \sum_{j=1}^{J}(b_j - \beta_j)^2(\sum_{k=1}^{K} x_{jk}^2 - K\overline{x}_j^2)
\end{aligned}$$

and that the random variables Y_{jk}, $\overline{Y}_j$ and b_j are all independent and have the following distributions

$$\begin{aligned}
Y_{jk} &\sim N(\alpha_j + \beta_j x_{jk}, \sigma^2), \\
\overline{Y}_j &\sim N(\alpha_j + \beta_j \overline{x}_j, \sigma^2/K), \\
b_j &\sim N(\beta_j, \sigma^2/(\sum_{k=1}^{K} x_{jk}^2 - K\overline{x}_j^2)).
\end{aligned}$$

2.4 Suppose you have the following data

x:	1.0	1.2	1.4	1.6	1.8	2.0
y:	3.15	4.85	6.50	7.20	8.25	16.50

and you want to fit a model with

$$\mathrm{E}(Y) = \ln(\beta_0 + \beta_1 x + \beta_2 x^2).$$

Write this model in the form of (2.13) specifying the vectors $\mathbf{y}$ and $\boldsymbol{\beta}$ and the matrix $\mathbf{X}$.

2.5 The model for two-factor analysis of variance with two levels of one factor, three levels of the other and no replication is

$$\mathrm{E}(Y_{jk}) = \mu_{jk} = \mu + \alpha_j + \beta_k; \qquad Y_{jk} \sim N(\mu_{jk}, \sigma^2)$$

where $j = 1, 2$; $k = 1, 2, 3$ and, using the sum-to-zero constraints, $\alpha_1 + \alpha_2 = 0, \beta_1 + \beta_2 + \beta_3 = 0$. Also the Y_{jk}'s are assumed to be independent. Write the equation for $\mathrm{E}(Y_{jk})$ in matrix notation. (Hint: let $\alpha_2 = -\alpha_1$, and $\beta_3 = -\beta_1 - \beta_2$).

3

Exponential Family and Generalized Linear Models

3.1 Introduction

Linear models of the form

$$E(Y_i) = \mu_i = \mathbf{x}_i^T \boldsymbol{\beta}; \qquad Y_i \sim N(\mu_i, \sigma^2) \tag{3.1}$$

where the random variables Y_i are independent are the basis of most analyses of continuous data. The transposed vector $\mathbf{x}_i^T$ represents the ith row of the design matrix $\mathbf{X}$. The example about the relationship between birthweight and gestational age is of this form, see Section 2.2.2. So is the exercise on plant growth where Y_i is the dry weight of plants and $\mathbf{X}$ has elements to identify the treatment and control groups (Exercise 2.1). Generalizations of these examples to the relationship between a continuous response and several explanatory variables (multiple regression) and comparisons of more than two means (analysis of variance) are also of this form.

Advances in statistical theory and computer software allow us to use methods analogous to those developed for linear models in the following more general situations:

1. Response variables have distributions other than the Normal distribution – they may even be categorical rather than continuous.
2. Relationship between the response and explanatory variables need not be of the simple linear form in (3.1).

One of these advances has been the recognition that many of the 'nice' properties of the Normal distribution are shared by a wider class of distributions called the **exponential family of distributions**. These distributions and their properties are discussed in the next section.

A second advance is the extension of the numerical methods to estimate the parameters $\boldsymbol{\beta}$ from the linear model described in (3.1) to the situation where there is some non-linear function relating $\mathrm{E}(Y_i) = \mu_i$ to the linear component $\mathbf{x}_i^T \boldsymbol{\beta}$, that is

$$g(\mu_i) = \mathbf{x}_i^T \boldsymbol{\beta}$$

(see Section 2.4). The function g is called the **link function**. In the initial formulation of generalized linear models by Nelder and Wedderburn (1972) and in most of the examples considered in this book, g is a simple mathematical function. These models have now been further generalized to situations where functions may be estimated numerically; such models are called **generalized additive models** (see Hastie and Tibshirani, 1990). In theory, the estimation is straightforward. In practice, it may require a considerable amount of computation involving numerical optimization of non-linear functions. Procedures to do these calculations are now included in many statistical programs.

This chapter introduces the exponential family of distributions and defines generalized linear models. Methods for parameter estimation and hypothesis testing are developed in Chapters 4 and 5, respectively.

3.2 Exponential family of distributions

Consider a single random variable Y whose probability distribution depends on a single parameter θ. The distribution belongs to the exponential family if it can be written in the form

$$f(y;\theta) = s(y)t(\theta)e^{a(y)b(\theta)} \tag{3.2}$$

where a, b, s and t are known functions. Notice the symmetry between y and θ. This is emphasized if equation (3.2) is rewritten as

$$f(y;\theta) = \exp[a(y)b(\theta) + c(\theta) + d(y)] \tag{3.3}$$

where $s(y) = \exp d(y)$ and $t(\theta) = \exp c(\theta)$.

If $a(y) = y$, the distribution is said to be in **canonical** (that is, standard) **form** and $b(\theta)$ is sometimes called the **natural parameter** of the distribution.

If there are other parameters, in addition to the parameter of interest θ, they are regarded as **nuisance parameters** forming parts of the functions a, b, c and d, and they are treated as though they are known.

Many well-known distributions belong to the exponential family. For example, the Poisson, Normal and binomial distributions can all be written in the canonical form – see Table 3.1.

3.2.1 Poisson distribution

The probability function for the discrete random variable Y is

$$f(y,\theta) = \frac{\theta^y e^{-\theta}}{y!}$$

Table 3.1 *Poisson, Normal and binomial distributions as members of the exponential family.*

Distribution	Natural parameter	c	d
Poisson	$\log\theta$	$-\theta$	$-\log y!$
Normal	$\frac{\mu}{\sigma^2}$	$-\frac{\mu^2}{2\sigma^2} - \frac{1}{2}\log(2\pi\sigma^2)$	$-\frac{y^2}{2\sigma^2}$
Binomial	$\log\left(\frac{\pi}{1-\pi}\right)$	$n\log(1-\pi)$	$\log\binom{n}{y}$

where y takes the values $0, 1, 2, \ldots$. This can be rewritten as

$$f(y, \theta) = \exp(y \log \theta - \theta - \log y!)$$

which is in the canonical form because $a(y) = y$. Also the natural parameter is $\log \theta$.

The Poisson distribution, denoted by $Y \sim Poisson(\theta)$, is used to model count data. Typically these are the number of occurrences of some event in a defined time period or space, when the probability of an event occurring in a very small time (or space) is low and the events occur independently. Examples include: the number of medical conditions reported by a person (Example 2.2.1), the number of tropical cyclones during a season (Example 1.6.4), the number of spelling mistakes on the page of a newspaper, or the number of faulty components in a computer or in a batch of manufactured items. If a random variable has the Poisson distribution, its expected value and variance are equal. Real data that might be plausibly modelled by the Poisson distribution often have a larger variance and are said to be **overdispersed,** and the model may have to be adapted to reflect this feature. Chapter 9 describes various models based on the Poisson distribution.

3.2.2 Normal distribution

The probability density function is

$$f(y; \mu) = \frac{1}{(2\pi\sigma^2)^{1/2}} \exp\left[-\frac{1}{2\sigma^2}(y - \mu)^2\right]$$

where μ is the parameter of interest and σ^2 is regarded as a nuisance parameter. This can be rewritten as

$$f(y; \mu) = \exp\left[-\frac{y^2}{2\sigma^2} + \frac{y\mu}{\sigma^2} - \frac{\mu^2}{2\sigma^2} - \frac{1}{2}\log(2\pi\sigma^2)\right].$$

This is in the canonical form. The natural parameter is $b(\mu) = \mu/\sigma^2$ and the other terms in (3.3) are

$$c(\mu) = -\frac{\mu^2}{2\sigma^2} - \frac{1}{2}\log(2\pi\sigma^2) \text{ and } d(y) = -\frac{y^2}{2\sigma^2}$$

(alternatively, the term $-\frac{1}{2}\log(2\pi\sigma^2)$ could be included in $d(y)$).

The Normal distribution is used to model continuous data that have a symmetric distribution. It is widely used for three main reasons. First, many naturally occurring phenomena are well described by the Normal distribution; for example, height or blood pressure of people. Second, even if data are not Normally distributed (e.g., if their distribution is skewed) the average or total of a random sample of values will be approximately Normally distributed; this result is proved in the Central Limit Theorem. Third, there is a great deal of statistical theory developed for the Normal distribution, including sampling distributions derived from it and approximations to other distributions. For these reasons, if continuous data $\mathbf{y}$ are not Normally distributed it is often worthwhile trying to identify a transformation, such as $y' = \log y$ or $y' = \sqrt{y}$, which produces data $\mathbf{y}'$ that are approximately Normal.

3.2.3 Binomial distribution

Consider a series of binary events, called 'trials', each with only two possible outcomes: 'success' or 'failure'. Let the random variable Y be the number of 'successes' in n independent trials in which the probability of success, π, is the same in all trials. Then Y has the binomial distribution with probability density function

$$f(y;\pi) = \binom{n}{y} \pi^y (1-\pi)^{n-y}$$

where y takes the values $0, 1, 2, \ldots, n$. This is denoted by $Y \sim binomial(n, \pi)$. Here π is the parameter of interest and n is assumed to be known. The probability function can be rewritten as

$$f(y;\mu) = \exp\left[y\log\pi - y\log(1-\pi) + n\log(1-\pi) + \log\binom{n}{y}\right]$$

which is of the form (3.3) with $b(\pi) = \log\pi - \log(1-\pi) = \log\left[\pi/(1-\pi)\right]$.

The binomial distribution is usually the model of first choice for observations of a process with binary outcomes. Examples include: the number of candidates who pass a test (the possible outcomes for each candidate being to pass or to fail), or the number of patients with some disease who are alive at a specified time since diagnosis (the possible outcomes being survival or death).

Other examples of distributions belonging to the exponential family are given in the exercises at the end of the chapter; not all of them are of the canonical form.

3.3 Properties of distributions in the exponential family

We need expressions for the expected value and variance of $a(Y)$. To find these we use the following results that apply for any probability density function provided that the order of integration and differentiation can be interchanged. From the definition of a probability density function, the area under the curve is unity so

$$\int f(y;\theta)\,dy = 1 \tag{3.4}$$

where integration is over all possible values of y. (If the random variable Y is discrete then integration is replaced by summation.)

If we differentiate both sides of (3.4) with respect to θ we obtain

$$\frac{d}{d\theta}\int f(y;\theta)dy = \frac{d}{d\theta}.1 = 0 \tag{3.5}$$

If the order of integration and differentiation in the first term is reversed then (3.5) becomes

$$\int \frac{df(y;\theta)}{d\theta} dy = 0 \tag{3.6}$$

Similarly if (3.4) is differentiated twice with respect to θ and the order of integration and differentiation is reversed we obtain

$$\int \frac{d^2 f(y;\theta)}{d\theta^2} dy = 0. \tag{3.7}$$

These results can now be used for distributions in the exponential family. From (3.3)

$$f(y;\theta) = \exp\left[a(y)b(\theta) + c(\theta) + d(y)\right]$$

so

$$\frac{df(y;\theta)}{d\theta} = \left[a(y)b'(\theta) + c'(\theta)\right] f(y;\theta).$$

By (3.6)

$$\int \left[a(y)b'(\theta) + c'(\theta)\right] f(y;\theta) dy = 0.$$

This can be simplified to

$$b'(\theta)\mathrm{E}[a(y)] + c'(\theta) = 0 \tag{3.8}$$

because $\int a(y)f(y;\theta)dy = \mathrm{E}[a(y)]$ by the definition of the expected value and $\int c'(\theta)f(y;\theta)dy = c'(\theta)$ by (3.4). Rearranging (3.8) gives

$$\mathrm{E}[a(Y)] = -c'(\theta)/b'(\theta). \tag{3.9}$$

A similar argument can be used to obtain $\mathrm{var}[a(Y)]$.

$$\frac{d^2 f(y;\theta)}{d\theta^2} = \left[a(y)b''(\theta) + c''(\theta)\right] f(y;\theta) + \left[a(y)b'(\theta) + c'(\theta)\right]^2 f(y;\theta) \tag{3.10}$$

The second term on the right hand side of (3.10) can be rewritten as

$$[b'(\theta)]^2\{a(y) - E[a(Y)]\}^2 f(y;\theta)$$

using (3.9). Then by (3.7)

$$\int \frac{d^2 f(y;\theta)}{d\theta^2} dy = b''(\theta)\mathrm{E}[a(Y)] + c''(\theta) + [b'(\theta)]^2\mathrm{var}[a(Y)] = 0 \tag{3.11}$$

because $\int \{a(y) - \mathrm{E}[a(Y)]\}^2 f(y;\theta)dy = \mathrm{var}[a(Y)]$ by definition.

Rearranging (3.11) and substituting (3.9) gives

$$\mathrm{var}[a(Y)] = \frac{b''(\theta)c'(\theta) - c''(\theta)b'(\theta)}{[b'(\theta)]^3} \tag{3.12}$$

Equations (3.9) and (3.12) can readily be verified for the Poisson, Normal and binomial distributions (see Exercise 3.4) and used to obtain the expected value and variance for other distributions in the exponential family.

We also need expressions for the expected value and variance of the derivatives of the log-likelihood function. From (3.3), the log-likelihood function for a distribution in the exponential family is

$$l(\theta; y) = a(y)b(\theta) + c(\theta) + d(y).$$

The derivative of $l(\theta; y)$ with respect to θ is

$$U(\theta; y) = \frac{dl(\theta; y)}{d\theta} = a(y)b'(\theta) + c'(\theta).$$

The function U is called the **score statistic** and, as it depends on y, it can be regarded as a random variable, that is

$$U = a(Y)b'(\theta) + c'(\theta). \quad (3.13)$$

Its expected value is

$$\mathrm{E}(U) = b'(\theta)\mathrm{E}[a(Y)] + c'(\theta).$$

From (3.9)

$$\mathrm{E}(U) = b'(\theta)\left[-\frac{c'(\theta)}{b'(\theta)}\right] + c'(\theta) = 0. \quad (3.14)$$

The variance of U is called the **information** and will be denoted by $\mathfrak{I}$. Using the formula for the variance of a linear transformation of random variables (see (1.3) and (3.13))

$$\mathfrak{I} = \mathrm{var}(U) = \left[b'(\theta)^2\right]\mathrm{var}[a(Y)].$$

Substituting (3.12) gives

$$\mathrm{var}(U) = \frac{b''(\theta)c'(\theta)}{b'(\theta)} - c''(\theta). \quad (3.15)$$

The score statistic U is used for inference about parameter values in generalized linear models (see Chapter 5).

Another property of U which will be used later is

$$\mathrm{var}(U) = \mathrm{E}(U^2) = -\mathrm{E}(U'). \quad (3.16)$$

The first equality follows from the general result

$$\mathrm{var}(X) = \mathrm{E}(X^2) - [\mathrm{E}(X)]^2$$

for any random variable, and the fact that $\mathrm{E}(U) = 0$ from (3.14). To obtain the second equality, we differentiate U with respect to θ; from (3.13)

$$U' = \frac{dU}{d\theta} = a(Y)b''(\theta) + c''(\theta).$$

Therefore the expected value of U' is

$$\begin{aligned}\mathrm{E}(U') &= b''(\theta)\mathrm{E}[a(Y)] + c''(\theta) \\ &= b''(\theta)\left[-\frac{c'(\theta)}{b'(\theta)}\right] + c''(\theta) \quad (3.17) \\ &= -\mathrm{var}(U) = -\mathfrak{I}\end{aligned}$$

by substituting (3.9) and then using (3.15).

3.4 Generalized linear models

The unity of many statistical methods was demonstrated by Nelder and Wedderburn (1972) using the idea of a generalized linear model. This model is defined in terms of a set of independent random variables $Y_1, \ldots, Y_N$ each with a distribution from the exponential family and the following properties:

1. The distribution of each Y_i has the canonical form and depends on a single parameter θ_i (the θ_i's do not all have to be the same), thus

$$f(y_i; \theta_i) = \exp\left[y_i b_i(\theta_i) + c_i(\theta_i) + d_i(y_i)\right].$$

2. The distributions of all the Y_i's are of the same form (e.g., all Normal or all binomial) so that the subscripts on b, c and d are not needed.

Thus the joint probability density function of $Y_1, \ldots, Y_N$ is

$$f(y_1, \ldots, y_N; \theta_1, \ldots, \theta_N) = \prod_{i=1}^{N} \exp\left[y_i b(\theta_i) + c(\theta_i) + d(y_i)\right] \qquad (3.18)$$

$$= \exp\left[\sum_{i=1}^{N} y_i b(\theta_i) + \sum_{i=1}^{N} c(\theta_i) + \sum_{i=1}^{N} d(y_i)\right]. \qquad (3.19)$$

The parameters θ_i are typically not of direct interest (since there may be one for each observation). For model specification we are usually interested in a smaller set of parameters $\beta_1, \ldots, \beta_p$ (where $p < N$). Suppose that $\mathrm{E}(Y_i) = \mu_i$ where μ_i is some function of θ_i. For a generalized linear model there is a transformation of μ_i such that

$$g(\mu_i) = \mathbf{x}_i^T \boldsymbol{\beta}.$$

In this equation

g is a monotone, differentiable function called the **link function**; $\mathbf{x}_i$ is a $p \times 1$ vector of explanatory variables (covariates and dummy variables for levels of factors),

$$\mathbf{x}_i = \begin{bmatrix} x_{i1} \\ \vdots \\ x_{ip} \end{bmatrix} \text{ so } \mathbf{x}_i^T = \begin{bmatrix} x_{i1} & \cdots & x_{ip} \end{bmatrix}$$

and $\boldsymbol{\beta}$ is the $p \times 1$ vector of parameters $\boldsymbol{\beta} = \begin{bmatrix} \beta_1 \\ \vdots \\ \beta_p \end{bmatrix}$. The vector $\mathbf{x}_i$ is the ith column of the design matrix $\mathbf{X}$.

Thus a generalized linear model has three components:

1. Response variables $Y_1, \ldots, Y_N$ which are assumed to share the same distribution from the exponential family;
2. A set of parameters $\boldsymbol{\beta}$ and explanatory variables

$$\mathbf{X} = \begin{bmatrix} \mathbf{x}_1^T \\ \vdots \\ \mathbf{x}_N^T \end{bmatrix} = \begin{bmatrix} x_{11} & \ldots & x_{1p} \\ \vdots & & \vdots \\ x_{N1} & & x_{Np} \end{bmatrix};$$

3. A monotone link function g such that

$$g(\mu_i) = \mathbf{x}_i^T \boldsymbol{\beta}$$

where

$$\mu_i = \mathrm{E}(Y_i).$$

This chapter concludes with three examples of generalized linear models.

3.5 Examples

3.5.1 Normal Linear Model

The best known special case of a generalized linear model is the model

$$\mathrm{E}(Y_i) = \mu_i = \mathbf{x}_i^T \boldsymbol{\beta}; \qquad Y_i \sim N(\mu_i, \sigma^2)$$

where $Y_1, ..., Y_N$ are independent. Here the link function is the identity function, $g(\mu_i) = \mu_i$. This model is usually written in the form

$$\mathbf{y} = \mathbf{X}\boldsymbol{\beta} + \mathbf{e}$$

where $\mathbf{e} = \begin{bmatrix} e_1 \\ \vdots \\ e_N \end{bmatrix}$ and the e_i's are independent, identically distributed random variables with $e_i \sim N(0, \sigma^2)$ for $i = 1, ..., N$.

In this form, the linear component $\boldsymbol{\mu} = \mathbf{X}\boldsymbol{\beta}$ represents the 'signal' and $\mathbf{e}$ represents the 'noise', random variation or 'error'. Multiple regression, analysis of variance and analysis of covariance are all of this form. These models are considered in Chapter 6.

3.5.2 Historical Linguistics

Consider a language which is the descendent of another language; for example, modern Greek is a descendent of ancient Greek, and the Romance languages are descendents of Latin. A simple model for the change in vocabulary is that if the languages are separated by time t then the probability that they have cognate words for a particular meaning is $e^{-\theta t}$ where θ is a parameter (see Figure 3.1). It is believed that θ is approximately the same for many commonly used meanings. For a test list of N different commonly used meanings suppose that a linguist judges, for each meaning, whether the corresponding words in

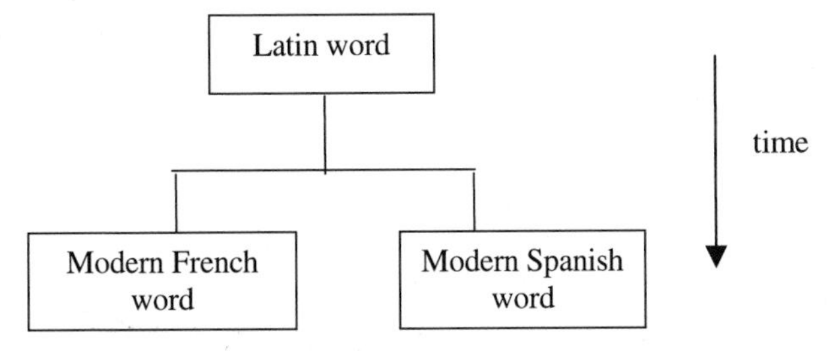

Figure 3.1 *Schematic diagram for the example on historical linguistics.*

two languages are cognate or not cognate. We can develop a generalized linear model to describe this situation.

Define random variables $Y_1, \ldots, Y_N$ as follows:

$$Y_i = \begin{cases} 1 & \text{if the languages have cognate words for meaning } i, \\ 0 & \text{if the words are not cognate.} \end{cases}$$

Then

$$P(Y_i = 1) = e^{-\theta t}$$

and

$$P(Y_i = 0) = 1 - e^{-\theta t}.$$

This is a special case of the distribution *binomial*(n, π) with $n = 1$ and $\mathrm{E}(Y_i) = \pi = e^{-\theta t}$. In this case the link function g is taken as logarithmic

$$g(\pi) = \log \pi = -\theta t$$

so that $g[\mathrm{E}(Y)]$ is linear in the parameter θ. In the notation used above, $\mathbf{x}_i = [-t]$ (the same for all i) and $\boldsymbol{\beta} = [\theta]$.

3.5.3 Mortality Rates

For a large population the probability of a randomly chosen individual dying at a particular time is small. If we assume that deaths from a non-infectious disease are independent events, then the number of deaths Y in a population can be modelled by a Poisson distribution

$$f(y; \mu) = \frac{\mu^y e^{-\mu}}{y!}$$

where y can take the values $0, 1, 2, \ldots$ and $\mu = \mathrm{E}(Y)$ is the expected number of deaths in a specified time period, such as a year.

The parameter μ will depend on the population size, the period of observation and various characteristics of the population (e.g., age, sex and medical history). It can be modelled, for example, by

$$\mathrm{E}(Y) = \mu = n\lambda(\mathbf{x}^T\boldsymbol{\beta})$$

Table 3.2 *Numbers of deaths from coronary heart disease and population sizes by 5-year age groups for men in the Hunter region of New South Wales, Australia in 1991.*

Age group (years)	Number of deaths, y_i	Population size, n_i	Rate per 100,000 men per year, $y_i/n_i \times 10,000$
30 - 34	1	17,742	5.6
35 - 39	5	16,554	30.2
40 - 44	5	16,059	31.1
45 - 49	12	13,083	91.7
50 - 54	25	10,784	231.8
55 - 59	38	9,645	394.0
60 - 64	54	10,706	504.4
65 - 69	65	9,933	654.4

Figure 3.2 *Death rate per 100,000 men (on a logarithmic scale) plotted against age.*

where n is the population size and $\lambda(\mathbf{x}^T\boldsymbol{\beta})$ is the rate per 100,000 people per year (which depends on the population characteristics described by the linear component $\mathbf{x}^T\boldsymbol{\beta}$).

Changes in mortality with age can be modelled by taking independent random variables $Y_1, \ldots, Y_N$ to be the numbers of deaths occurring in successive age groups. For example, Table 3.2 shows age-specific data for deaths from coronary heart disease.

Figure 3.2 shows how the mortality rate $y_i/n_i \times 100,000$ increases with age. Note that a logarithmic scale has been used on the vertical axis. On this scale the scatter plot is approximately linear, suggesting that the relationship between y_i/n_i and age group i is approximately exponential. Therefore a possible model is

$$\mathrm{E}(Y_i) = \mu_i = n_i e^{\theta i} \quad ; \quad Y_i \sim Poisson(\mu_i),$$

where $i = 1$ for the age group 30-34 years, $i = 2$ for 35-39, ..., $i = 8$ for 65-69 years.

This can be written as a generalized linear model using the logarithmic link function

$$g(\mu_i) = \log \mu_i = \log n_i + \theta i$$

which has the linear component $\mathbf{x}_i^T \boldsymbol{\beta}$ with $\mathbf{x}_i^T = \begin{bmatrix} \log n_i & i \end{bmatrix}$ and $\boldsymbol{\beta} = \begin{bmatrix} 1 \\ \theta \end{bmatrix}$.

3.6 Exercises

3.1 The following relationships can be described by generalized linear models. For each one, identify the response variable and the explanatory variables, select a probability distribution for the response (justifying your choice) and write down the linear component.

(a) The effect of age, sex, height, mean daily food intake and mean daily energy expenditure on a person's weight.

(b) The proportions of laboratory mice that became infected after exposure to bacteria when five different exposure levels are used and 20 mice are exposed at each level.

(c) The relationship between the number of trips per week to the supermarket for a household and the number of people in the household, the household income and the distance to the supermarket.

3.2 If the random variable Y has the **Gamma distribution** with a scale parameter θ, which is the parameter of interest, and a known shape parameter ϕ, then its probability density function is

$$f(y;\theta) = \frac{y^{\phi-1}\theta^{\phi}e^{-y\theta}}{\Gamma(\phi)}.$$

Show that this distribution belongs to the exponential family and find the natural parameter. Also using results in this chapter, find $\mathrm{E}(Y)$ and $\mathrm{var}(Y)$.

3.3 Show that the following probability density functions belong to the exponential family:

(a) Pareto distribution $f(y;\theta) = \theta y^{-\theta-1}$.

(b) Exponential distribution $f(y;\theta) = \theta e^{-y\theta}$.

(c) Negative binomial distribution

$$f(y;\theta) = \binom{y+r-1}{r-1} \theta^r (1-\theta)^y$$

where r is known.

3.4 Use results (3.9) and (3.12) to verify the following results:

(a) For $Y \sim Poisson(\theta)$, $\mathrm{E}(Y) = \mathrm{var}(Y) = \theta$.
(b) For $Y \sim N(\mu, \sigma^2)$, $\mathrm{E}(Y) = \mu$ and $\mathrm{var}(Y) = \sigma^2$.
(c) For $Y \sim binomial(n, \pi)$, $\mathrm{E}(Y) = n\pi$ and $\mathrm{var}(Y) = n\pi(1-\pi)$.

3.5 Do you consider the model suggested in Example 3.5.3 to be adequate for the data shown in Figure 3.2? Justify your answer. Use simple linear regression (with suitable transformations of the variables) to obtain a model for the change of death rates with age. How well does the model fit the data? (Hint: compare observed and expected numbers of deaths in each groups.)

3.6 Consider N independent binary random variables $Y_1, \ldots, Y_N$ with

$$P(Y_i = 1) = \pi_i \text{ and } P(Y_i = 0) = 1 - \pi_i \, .$$

The probability function of Y_i can be written as

$$\pi_i^{y_i} (1 - \pi_i)^{1-y_i}$$

where $y_i = 0$ or 1.

(a) Show that this probability function belongs to the exponential family of distributions.
(b) Show that the natural parameter is

$$\log \left(\frac{\pi_i}{1 - \pi_i} \right).$$

This function, the logarithm of the **odds** $\pi_i/(1 - \pi_i)$, is called the **logit** function.

(c) Show that $\mathrm{E}(Y_i) = \pi_i$.
(d) If the link function is

$$g(\pi) = \log \left(\frac{\pi}{1 - \pi} \right) = \mathbf{x}^T \boldsymbol{\beta}$$

show that this is equivalent to modelling the probability π as

$$\pi = \frac{e^{\mathbf{x}^T \boldsymbol{\beta}}}{1 + e^{\mathbf{x}^T \boldsymbol{\beta}}}.$$

(e) In the particular case where $\mathbf{x}^T \boldsymbol{\beta} = \beta_1 + \beta_2 x$, this gives

$$\pi = \frac{e^{\beta_1 + \beta_2 x}}{1 + e^{\beta_1 + \beta_2 x}}$$

which is the **logistic function**.

(f) Sketch the graph of π against x in this case, taking β_1 and β_2 as constants. How would you interpret this graph if x is the dose of an insecticide and π is the probability of an insect dying?

3.7 Is the **extreme value (Gumbel) distribution**, with probability density function

$$f(y;\theta) = \frac{1}{\phi} \exp\left\{ \frac{(y-\theta)}{\phi} - \exp\left[\frac{(y-\theta)}{\phi} \right] \right\}$$

(where $\phi > 0$ regarded as a nuisance parameter) a member of the exponential family?

3.8 Suppose $Y_1, ..., Y_N$ are independent random variables each with the Pareto distribution and

$$\mathrm{E}(Y_i) = (\beta_0 + \beta_1 x_i)^2.$$

Is this a generalized linear model? Give reasons for your answer.

3.9 Let $Y_1, \ldots, Y_N$ be independent random variables with

$$\mathrm{E}(Y_i) = \mu_i = \beta_0 + \log(\beta_1 + \beta_2 x_i); \quad Y_i \sim N(\mu, \sigma^2)$$

for all $i = 1, ..., N$. Is this a generalized linear model? Give reasons for your answer.

3.10 For the Pareto distribution find the score statistics U and the information $\mathfrak{I} = \mathrm{var}(U)$. Verify that $\mathrm{E}(U) = 0$.

4

Estimation

4.1 Introduction

This chapter is about obtaining point and interval estimates of parameters for generalized linear models using methods based on maximum likelihood. Although explicit mathematical expressions can be found for estimators in some special cases, numerical methods are usually needed. Typically these methods are iterative and are based on the Newton-Raphson algorithm. To illustrate this principle, the chapter begins with a numerical example. Then the theory of estimation for generalized linear models is developed. Finally there is another numerical example to demonstrate the methods in detail.

4.2 Example: Failure times for pressure vessels

The data in Table 4.1 are the lifetimes (times to failure in hours) of Kevlar epoxy strand pressure vessels at 70% stress level. They are given in Table 29.1 of the book of data sets by Andrews and Herzberg (1985).

Figure 4.1 shows the shape of their distribution.

A commonly used model for times to failure (or survival times) is the **Weibull distribution** which has the probability density function

$$f(y;\lambda,\theta)=\frac{\lambda y^{\lambda-1}}{\theta^{\lambda}}\exp\left[-\left(\frac{y}{\theta}\right)^{\lambda}\right] \tag{4.1}$$

where $y>0$ is the time to failure, λ is a parameter that determines the shape of the distribution and θ is a parameter that determines the scale. Figure 4.2 is a probability plot of the data in Table 4.1 compared to the Weibull distribution with $\lambda=2$. Although there are discrepancies between the distribution and the data for some of the shorter times, for most of the

Table 4.1 *Lifetimes of pressure vessels.*

1051	4921	7886	10861	13520
1337	5445	8108	11026	13670
1389	5620	8546	11214	14110
1921	5817	8666	11362	14496
1942	5905	8831	11604	15395
2322	5956	9106	11608	16179
3629	6068	9711	11745	17092
4006	6121	9806	11762	17568
4012	6473	10205	11895	17568
4063	7501	10396	12044	

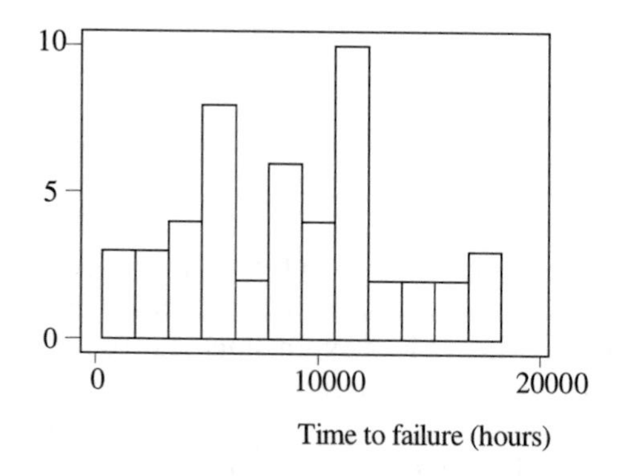

Figure 4.1 *Distribution of lifetimes of pressure vessels.*

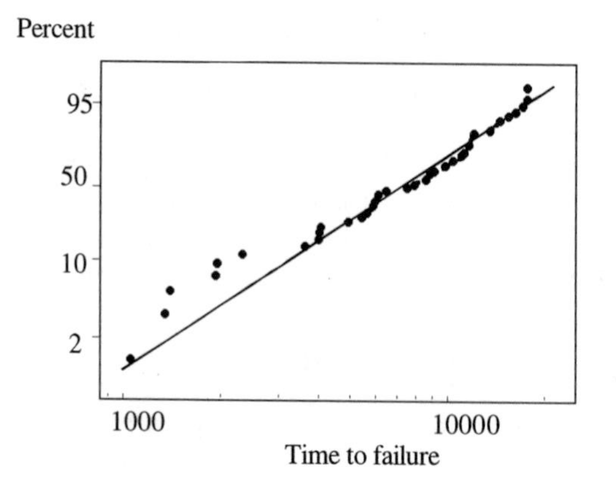

Figure 4.2 *Probability plot of the data on lifetimes of pressure vessels compared to the Weibull distribution with shape parameter = 2.*

observations the distribution appears to provide a good model for the data. Therefore we will use a Weibull distribution with $\lambda = 2$ and estimate θ.

The distribution in (4.1) can be written as

$$f(y;\theta) = \exp\left[\log\lambda + (\lambda-1)\log y - \lambda\log\theta - (y/\theta)^{\lambda}\right].$$

This belongs to the exponential family (3.2) with

$$a(y) = y^{\lambda}, b(\theta) = -\theta^{-\lambda}, c(\theta) = \log\lambda - \lambda\log\theta \text{ and } d(y) = (\lambda-1)\log y \quad (4.2)$$

where λ is a nuisance parameter. This is not in the canonical form (unless $\lambda = 1$, corresponding to the exponential distribution) and so it cannot be used directly in the specification of a generalized linear model. However it is

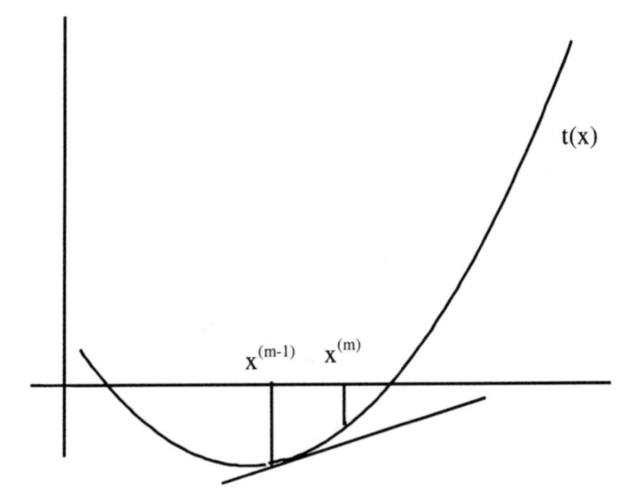

Figure 4.3 *Newton-Raphson method for finding the solution of the equation t(x)=0.*

suitable for illustrating the estimation of parameters for distributions in the exponential family.

Let $Y_1, ..., Y_N$ denote the data, with $N = 49$. If the data are from a random sample of pressure vessels, we assume the Y_i's are independent random variables. If they all have the Weibull distribution with the same parameters, their joint probability distribution is

$$f(y_1, ..., y_N; \theta, \lambda) = \prod_{i=1}^{N} \frac{\lambda y_i^{\lambda-1}}{\theta^\lambda} \exp\left[-\left(\frac{y_i}{\theta}\right)^\lambda\right].$$

The log-likelihood function is

$$f(\theta; y_1, ..., y_N, \lambda) = \sum_{i=1}^{N} \left[[(\lambda - 1)\log y_i + \log \lambda - \lambda \log \theta] - \left(\frac{y_i}{\theta}\right)^\lambda\right]. \quad (4.3)$$

To maximize this function we require the derivative with respect to θ. This is the score function

$$\frac{dl}{d\theta} = U = \sum_{i=1}^{N} \left[\frac{-\lambda}{\theta} + \frac{\lambda y_i^\lambda}{\theta^{\lambda+1}}\right] \quad (4.4)$$

The maximum likelihood estimator $\widehat{\theta}$ is the solution of the equation $U(\theta) = 0$. In this case it is easy to find an explicit expression for $\widehat{\theta}$ if λ is a known constant, but for illustrative purposes, we will obtain a numerical solution using the Newton-Raphson approximation.

Figure 4.3 shows the principle of the Newton-Raphson algorithm. We want to find the value of x at which the function t crosses the x-axis, i.e., where

$t(x) = 0$. The slope of t at a value $x^{(m-1)}$ is given by

$$\left[\frac{dt}{dx}\right]_{x=x^{(m-1)}} = t'(x^{(m-1)}) = \frac{t(x^{(m)}) - t(x^{(m-1)})}{x^{(m)} - x^{(m-1)}} \tag{4.5}$$

where the distance $x^{(m)} - x^{(m-1)}$ is small. If $x^{(m)}$ is the required solution so that $t\,(x^m) = 0$, then (4.5) can be re-arranged to give

$$x^{(m)} = x^{(m-1)} - \frac{t(x^{(m-1)})}{t'(x^{(m-1)})}. \tag{4.6}$$

This is the Newton-Raphson formula for solving $t(x) = 0$. Starting with an initial guess $x^{(1)}$ successive approximations are obtained using (4.6) until the iterative process converges.

For maximum likelihood estimation using the score function, the estimating equation equivalent to (4.6) is

$$\theta^{(m)} = \theta^{(m-1)} - \frac{U^{(m-1)}}{U'^{(m-1)}}. \tag{4.7}$$

From (4.4), for the Weibull distribution with $\lambda = 2$,

$$U = -\frac{2 \times N}{\theta} + \frac{2 \times \sum y_i^2}{\theta^3} \tag{4.8}$$

which is evaluated at successive estimates $\theta^{(m)}$. The derivative of U, obtained by differentiating (4.4), is

$$\begin{aligned} \frac{dU}{d\theta} = U' &= \sum_{i=1}^{N} \left[\frac{\lambda}{\theta^2} - \frac{\lambda(\lambda+1)y_i^\lambda}{\theta^{\lambda+2}}\right] \\ &= \frac{2 \times N}{\theta^2} - \frac{2 \times 3 \times \sum y_i^2}{\theta^4}. \end{aligned} \tag{4.9}$$

For maximum likelihood estimation, it is common to approximate U' by its expected value $E(U')$. For distributions in the exponential family, this is readily obtained using expression (3.17). The information $\mathfrak{I}$ is

$$\begin{aligned} \mathfrak{I} = E(-U') &= E\left[-\sum_{i=1}^{N} U_i'\right] = \sum_{i=1}^{N} \left[E(-U_i')\right] \\ &= \sum_{i=1}^{N} \left[\frac{b''(\theta)c'(\theta)}{b'(\theta)} - c''(\theta)\right] \\ &= \frac{\lambda^2 N}{\theta^2} \end{aligned} \tag{4.10}$$

where U_i is the score for Y_i and expressions for b and c are given in (4.2). Thus an alternative estimating equation is

$$\theta^{(m)} = \theta^{(m-1)} + \frac{U^{(m-1)}}{\mathfrak{I}^{(m-1)}} \tag{4.11}$$

Table 4.2 *Details of Newton-Raphson iterations to obtain a maximum likelihood estimate for the scale parameter for the Weibull distribution to model the data in Table 4.1.*

Iteration	1	2	3	4
θ	8805.9	9633.9	9876.4	9892.1
$U \times 10^6$	2915.10	552.80	31.78	0.21
$U' \times 10^6$	-3.52	-2.28	-2.02	-2.00
$\mathrm{E}(U') \times 10^6$	-2.53	-2.11	-2.01	-2.00
U/U'	-827.98	-242.46	-15.73	-0.105
$U/\mathrm{E}(U')$	-1152.21	-261.99	-15.81	-0.105

This is called the **method of scoring**.

Table 4.2 shows the results of using equation (4.7) iteratively taking the mean of the data in Table 4.1, $\overline{y} = 8805.9$, as the initial value $\theta^{(1)}$; this and subsequent approximations are shown in the top row of Table 4.2. Numbers in the second row were obtained by evaluating (4.8) at $\theta^{(m)}$ and the data values; they approach zero rapidly. The third and fourth rows, U' and $\mathrm{E}(U') = -\mathfrak{I}$, have similar values illustrating that either could be used; this is further shown by the similarity of the numbers in the fifth and sixth rows. The final estimate is $\theta^{(5)} = 9892.1-(-0.105) = 9892.2$ – this is the maximum likelihood estimate $\widehat{\theta}$ for these data. At this value the log-likelihood function, calculated from (4.3), is $l = -480.850$.

Figure 4.4 shows the log-likelihood function for these data and the Weibull distribution with $\lambda = 2$. The maximum value is at $\widehat{\theta} = 9892.2$. The curvature of the function in the vicinity of the maximum determines the reliability of $\widehat{\theta}$. The curvature of l is defined by the rate of change of U, that is, by U'. If U', or $\mathrm{E}(U')$, is small then l is flat so that U is approximately zero for a wide interval of θ values. In this case $\widehat{\theta}$ is not well-determined and its standard error is large. In fact, it is shown in Chapter 5 that the variance of $\widehat{\theta}$ is inversely related to $\mathfrak{I} = \mathrm{E}(-U')$ and the standard error of $\widehat{\theta}$ is approximately

$$s.e.(\widehat{\theta}) = \sqrt{1/\mathfrak{I}}. \tag{4.12}$$

For this example, at $\widehat{\theta} = 9892.2, \mathfrak{I} = -\mathrm{E}(U') = 2.00 \times 10^{-6}$ so $s.e.(\widehat{\theta}) = 1/\sqrt{0.000002} = 707$. If the sampling distribution of $\widehat{\theta}$ is approximately Normal, a 95% confidence interval for θ is given approximately by

$$9892 \pm 1.96 \times 707, \text{ or } (8506, 11278).$$

The methods illustrated in this example are now developed for generalized linear models.

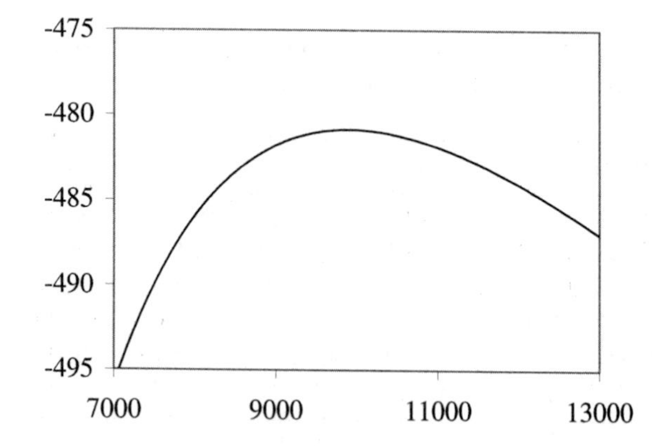

Figure 4.4 *Log-likelihood function for the pressure vessel data in Table 4.1.*

4.3 Maximum likelihood estimation

Consider independent random variables $Y_1, ..., Y_N$ satisfying the properties of a generalized linear model. We wish to estimate parameters $\boldsymbol{\beta}$ which are related to the Y_i's through $\mathrm{E}(Y_i) = \mu_i$ and $g(\mu_i) = \mathbf{x}_i^T \boldsymbol{\beta}$.

For each Y_i, the log-likelihood function is

$$l_i = y_i b(\theta_i) + c(\theta_i) + d(y_i) \tag{4.13}$$

where the functions b, c and d are defined in (3.3). Also

$$\mathrm{E}(Y_i) = \mu_i = -c'(\theta_i)/b'(\theta_i) \tag{4.14}$$

$$\mathrm{var}(Y_i) = \left[b''(\theta_i)c'(\theta_i) - c''(\theta_i)b'(\theta_i)\right] / \left[b'(\theta_i)\right]^3 \tag{4.15}$$

$$\text{and } g(\mu_i) = \mathbf{x}_i^T \boldsymbol{\beta} = \eta_i \tag{4.16}$$

where $\mathbf{x}_i$ is a vector with elements $x_{ij}, j = 1, ...p$.

The log-likelihood function for all the Y_i's is

$$l = \sum_{i=1}^{N} l_i = \sum y_i b(\theta_i) + \sum c(\theta_i) + \sum d(y_i).$$

To obtain the maximum likelihood estimator for the parameter β_j we need

$$\frac{\partial l}{\partial \beta_j} = U_j = \sum_{i=1}^{N} \left[\frac{\partial l_i}{\partial \beta_j}\right] = \sum_{i=1}^{N} \left[\frac{\partial l_i}{\partial \theta_i} . \frac{\partial \theta_i}{\partial \mu_i} . \frac{\partial \mu_i}{\partial \beta_j}\right] \tag{4.17}$$

using the chain rule for differentiation. We will consider each term on the right hand side of (4.17) separately. First

$$\frac{\partial l_i}{\partial \theta_i} = y_i b'(\theta_i) + c'(\theta_i) = b'(\theta_i)(y_i - \mu_i)$$

by differentiating (4.13) and substituting (4.14). Next

$$\frac{\partial \theta_i}{\partial \mu_i} = 1 \bigg/ \left(\frac{\partial \mu_i}{\partial \theta_i}\right).$$

Differentiation of (4.14) gives

$$\begin{aligned}\frac{\partial \mu_i}{\partial \theta_i} &= \frac{-c''(\theta_i)}{b'(\theta_i)} + \frac{c'(\theta_i)b''(\theta_i)}{[b'(\theta_i)]^2} \\ &= b'(\theta_i)\text{var}(Y_i)\end{aligned}$$

from (4.15). Finally, from (4.16)

$$\frac{\partial \mu_i}{\partial \beta_j} = \frac{\partial \mu_i}{\partial \eta_i} \cdot \frac{\partial \eta_i}{\partial \beta_j} = \frac{\partial \mu_i}{\partial \eta_i} x_{ij}.$$

Hence the score, given in (4.17), is

$$U_j = \sum_{i=1}^{N} \left[\frac{(y_i - \mu_i)}{\text{var}(Y_i)} x_{ij} \left(\frac{\partial \mu_i}{\partial \eta_i}\right)\right]. \tag{4.18}$$

The variance-covariance matrix of the U_j's has terms

$$\mathfrak{I}_{jk} = E\left[U_j U_k\right]$$

which form the **information matrix** $\mathfrak{I}$. From (4.18)

$$\begin{aligned}\mathfrak{I}_{jk} &= \text{E}\left\{\sum_{i=1}^{N}\left[\frac{(Y_i - \mu_i)}{\text{var}(Y_i)} x_{ij}\left(\frac{\partial \mu_i}{\partial \eta_i}\right)\right]\sum_{l=1}^{N}\left[\frac{(Y_l - \mu_l)}{\text{var}(Y_l)} x_{lk}\left(\frac{\partial \mu_l}{\partial \eta_l}\right)\right]\right\} \\ &= \sum_{i=1}^{N} \frac{\text{E}\left[(Y_i - \mu_i)^2\right] x_{ij} x_{ik}}{\left[\text{var}(Y_i)\right]^2}\left(\frac{\partial \mu_i}{\partial \eta_i}\right)^2 \end{aligned} \tag{4.19}$$

because $\text{E}[(Y_i - \mu_i)(Y_l - \mu_l)] = 0$ for $i \neq l$ as the Y_i's are independent. Using $\text{E}\left[(Y_i - \mu_i)^2\right] = \text{var}(Y_i)$, (4.19) can be simplified to

$$\mathfrak{I}_{jk} = \sum_{i=1}^{N} \frac{x_{ij} x_{ik}}{\text{var}(Y_i)}\left(\frac{\partial \mu_i}{\partial \eta_i}\right)^2. \tag{4.20}$$

The estimating equation (4.11) for the method of scoring generalizes to

$$\mathbf{b}^{(m)} = \mathbf{b}^{(m-1)} + \left[\mathfrak{I}^{(m-1)}\right]^{-1} \mathbf{U}^{(m-1)} \tag{4.21}$$

where $\mathbf{b}^{(m)}$ is the vector of estimates of the parameters $\beta_1, ..., \beta_p$ at the mth iteration. In equation (4.21), $\left[\mathfrak{I}^{(m-1)}\right]^{-1}$ is the inverse of the information matrix with elements $\mathfrak{I}_{jk}$ given by (4.20) and $\mathbf{U}^{(m-1)}$ is the vector of elements given by (4.18), all evaluated at $\mathbf{b}^{(m-1)}$. If both sides of equation (4.21) are multiplied by $\mathfrak{I}^{(m-1)}$ we obtain

$$\mathfrak{I}^{(m-1)}\mathbf{b}^{(m)} = \mathfrak{I}^{(m-1)}\mathbf{b}^{(m-1)} + \mathbf{U}^{(m-1)}. \tag{4.22}$$

From (4.20) $\mathfrak{I}$ can be written as

$$\mathfrak{I} = \mathbf{X}^T\mathbf{W}\mathbf{X}$$

where $\mathbf{W}$ is the $N \times N$ diagonal matrix with elements

$$w_{ii} = \frac{1}{\text{var}(Y_i)}\left(\frac{\partial \mu_i}{\partial \eta_i}\right)^2. \tag{4.23}$$

The expression on the right-hand side of (4.22) is the vector with elements

$$\sum_{k=1}^{p}\sum_{i=1}^{N}\frac{x_{ij}x_{ik}}{\text{var}(Y_i)}\left(\frac{\partial \mu_i}{\partial \eta_i}\right)^2 b_k^{(m-1)} + \sum_{i=1}^{N}\frac{(y_i-\mu_i)x_{ij}}{\text{var}(Y_i)}\left(\frac{\partial \mu_i}{\partial \eta_i}\right)$$

evaluated at $\mathbf{b}^{(m-1)}$; this follows from equations (4.20) and (4.18). Thus the right-hand side of equation (4.22) can be written as

$$\mathbf{X}^T\mathbf{W}\mathbf{z}$$

where $\mathbf{z}$ has elements

$$z_i = \sum_{k=1}^{p} x_{ik}b_k^{(m-1)} + (y_i-\mu_i)\left(\frac{\partial \eta_i}{\partial \mu_i}\right) \tag{4.24}$$

with μ_i and $\partial\eta_i/\partial\mu_i$ evaluated at $\mathbf{b}^{(m-1)}$.

Hence the iterative equation (4.22), can be written as

$$\mathbf{X}^T\mathbf{W}\mathbf{X}\mathbf{b}^{(m)} = \mathbf{X}^T\mathbf{W}\mathbf{z}. \tag{4.25}$$

This is the same form as the normal equations for a linear model obtained by weighted least squares, except that it has to be solved iteratively because, in general, $\mathbf{z}$ and $\mathbf{W}$ depend on $\mathbf{b}$. Thus for generalized linear models, maximum likelihood estimators are obtained by an **iterative weighted least squares** procedure (Charnes et al., 1976).

Most statistical packages that include procedures for fitting generalized linear models have an efficient algorithm based on (4.25). They begin by using some initial approximation $\mathbf{b}^{(0)}$ to evaluate $\mathbf{z}$ and $\mathbf{W}$, then (4.25) is solved to give $\mathbf{b}^{(1)}$ which in turn is used to obtain better approximations for $\mathbf{z}$ and $\mathbf{W}$, and so on until adequate convergence is achieved. When the difference between successive approximations $\mathbf{b}^{(m-1)}$ and $\mathbf{b}^{(m)}$ is sufficiently small, $\mathbf{b}^{(m)}$ is taken as the maximum likelihood estimate.

The example below illustrates the use of this estimation procedure.

4.4 Poisson regression example

The artificial data in Table 4.3 are counts y observed at various values of a covariate x. They are plotted in Figure 4.5.

Let us assume that the responses Y_i are Poisson random variables. In practice, such an assumption would be made either on substantive grounds or from noticing that in Figure 4.5 the variability increases with Y. This observation

Table 4.3 *Data for Poisson regression example.*

y_i	2	3	6	7	8	9	10	12	15
x_i	-1	-1	0	0	0	0	1	1	1

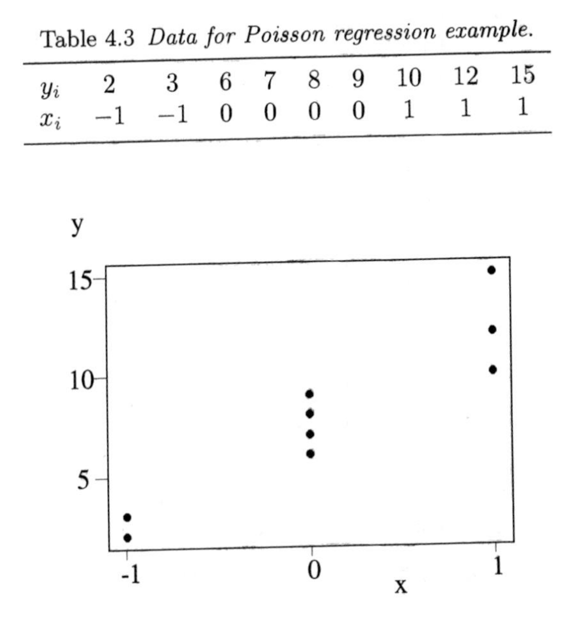

Figure 4.5 *Poisson regression example (data in Table 4.3).*

supports the use of the Poisson distribution which has the property that the expected value and variance of Y_i are equal

$$\mathrm{E}(Y_i) = \mathrm{var}(Y_i). \tag{4.26}$$

Let us model the relationship between Y_i and x_i by the straight line

$$\begin{aligned} \mathrm{E}(Y_i) = \mu_i &= \beta_1 + \beta_2 x_i \\ &= \mathbf{x}_i^T \boldsymbol{\beta} \end{aligned}$$

where

$$\boldsymbol{\beta} = \begin{bmatrix} \beta_1 \\ \beta_2 \end{bmatrix} \text{ and } \mathbf{x}_i = \begin{bmatrix} 1 \\ x_i \end{bmatrix}$$

for $i = 1, ..., N$. Thus we take the link function $g(\mu_i)$ to be the identity function

$$g(\mu_i) = \mu_i = \mathbf{x}_i^T \boldsymbol{\beta} = \eta_i.$$

Therefore $\partial\mu_i/\partial\eta_i = 1$ which simplifies equations (4.23) and (4.24). From (4.23) and (4.26)

$$w_{ii} = \frac{1}{\mathrm{var}(Y_i)} = \frac{1}{\beta_1 + \beta_2 x_i}.$$

Using the estimate $\mathbf{b} = \begin{bmatrix} b_1 \\ b_2 \end{bmatrix}$ for $\boldsymbol{\beta}$, equation (4.24) becomes

$$z_i = b_1 + b_2 x_i + (y_i - b_1 - b_2 x_i) = y_i.$$

Also

$$\mathfrak{I} = \mathbf{X}^T\mathbf{W}\mathbf{X} = \begin{bmatrix} \sum_{i=1}^N \dfrac{1}{b_1 + b_2 x_i} & \sum_{i=1}^N \dfrac{x_i}{b_1 + b_2 x_i} \\ \sum_{i=1}^N \dfrac{x_i}{b_1 + b_2 x_i} & \sum_{i=1}^N \dfrac{x_i^2}{b_1 + b_2 x_i} \end{bmatrix}$$

and

$$\mathbf{X}^T\mathbf{W}\mathbf{z} = \begin{bmatrix} \sum_{i=1}^N \dfrac{y_i}{b_1 + b_2 x_i} \\ \sum_{i=1}^N \dfrac{x_i y_i}{b_1 + b_2 x_i} \end{bmatrix}.$$

The maximum likelihood estimates are obtained iteratively from the equations

$$(\mathbf{X}^T\mathbf{W}\mathbf{X})^{(m-1)}\mathbf{b}^{(m)} = \mathbf{X}^T\mathbf{W}\mathbf{z}^{(m-1)}$$

where the superscript $^{(m-1)}$ denotes evaluation at $\mathbf{b}^{(m-1)}$.

For these data, $N = 9$

$$\mathbf{y} = \mathbf{z} = \begin{bmatrix} 2 \\ 3 \\ \vdots \\ 15 \end{bmatrix} \quad \text{and} \quad \mathbf{X} = \begin{bmatrix} \mathbf{x}_1 \\ \mathbf{x}_2 \\ \vdots \\ \mathbf{x}_9 \end{bmatrix} = \begin{bmatrix} 1 & -1 \\ 1 & -1 \\ \vdots & \vdots \\ 1 & 1 \end{bmatrix}.$$

From Figure 4.5 we can obtain initial estimates $b_1^{(1)} = 7$ and $b_2^{(1)} = 5$. Therefore

$$(\mathbf{X}^T\mathbf{W}\mathbf{X})^{(1)} = \begin{bmatrix} 1.821429 & -0.75 \\ -0.75 & 1.25 \end{bmatrix}, \qquad (\mathbf{X}^T\mathbf{W}\mathbf{z})^{(1)} = \begin{bmatrix} 9.869048 \\ 0.583333 \end{bmatrix}$$

$$\text{so} \quad \mathbf{b}^{(2)} = \left[(\mathbf{X}^T\mathbf{W}\mathbf{X})^{(1)}\right]^{-1}(\mathbf{X}^T\mathbf{W}\mathbf{z})^{(1)}$$

$$= \begin{bmatrix} 0.729167 & 0.4375 \\ 0.4375 & 1.0625 \end{bmatrix}\begin{bmatrix} 9.869048 \\ 0.583333 \end{bmatrix}$$

$$= \begin{bmatrix} 7.4514 \\ 4.9375 \end{bmatrix}.$$

This iterative process is continued until it converges. The results are shown in Table 4.4.

The maximum likelihood estimates are $\widehat{\beta}_1 = 7.45163$ and $\widehat{\beta}_2 = 4.93530$. At these values the inverse of the information matrix $\mathfrak{I} = \mathbf{X}^T\mathbf{W}\mathbf{X}$ is

$$\mathfrak{I}^{-1} = \begin{bmatrix} 0.7817 & 0.4166 \\ 0.4166 & 1.1863 \end{bmatrix}$$

(this is the variance-covariance matrix for $\widehat{\boldsymbol{\beta}}$ – see Section 5.4). So, for example,

Table 4.4 *Successive approximations for regression coefficients in the Poisson regression example.*

m	1	2	3	4
$b_1^{(m)}$	7	7.45139	7.45163	7.45163
$b_2^{(m)}$	5	4.93750	4.93531	4.93530

Table 4.5 *Numbers of cases of AIDS in Australia for successive quarter from 1984 to 1988.*

	Quarter			
Year	1	2	3	4
1984	1	6	16	23
1985	27	39	31	30
1986	43	51	63	70
1987	88	97	91	104
1988	110	113	149	159

an approximate 95% confidence interval for the slope β_2 is

$$4.9353 \pm 1.96\sqrt{1.1863} \text{ or } (2.80, 7.07).$$

4.5 Exercises

4.1 The data in Table 4.5 show the numbers of cases of AIDS in Australia by date of diagnosis for successive 3-months periods from 1984 to 1988. (Data from National Centre for HIV Epidemiology and Clinical Research, 1994.)

In this early phase of the epidemic, the numbers of cases seemed to be increasing exponentially.

(a) Plot the number of cases y_i against time period i $(i = 1, .., 20)$.

(b) A possible model is the Poisson distribution with parameter $\lambda_i = i^\theta$, or equivalently

$$\log \lambda_i = \theta \log i.$$

Plot $\log y_i$ against $\log i$ to examine this model.

(c) Fit a generalized linear model to these data using the Poisson distribution, the log-link function and the equation

$$g(\lambda_i) = \log \lambda_i = \beta_1 + \beta_2 x_i,$$

where $x_i = \log i$. Firstly, do this from first principles, working out expressions for the weight matrix $\mathbf{W}$ and other terms needed for the iterative equation

$$\mathbf{X}^T\mathbf{W}\mathbf{X}\mathbf{b}^{(m)} = \mathbf{X}^T\mathbf{W}\mathbf{z}$$

Table 4.6 *Survival time, y_i, in weeks and $\log_{10}$(initial white blood cell count), x_i, for seventeen leukemia patients.*

x_i	65	156	100	134	16	108	121	4	39
y_i	3.36	2.88	3.63	3.41	3.78	4.02	4.00	4.23	3.73
x_i	143	56	26	22	1	1	5	65	
y_i	3.85	3.97	4.51	4.54	5.00	5.00	4.72	5.00	

and using software which can perform matrix operations to carry out the calculations.

(d) Fit the model described in (c) using statistical software which can perform Poisson regression. Compare the results with those obtained in (c).

4.2 The data in Table 4.6 are times to death, y_i, in weeks from diagnosis and $\log_{10}$(initial white blood cell count), x_i, for seventeen patients suffering from leukemia. (This is Example U from Cox and Snell, 1981).

(a) Plot y_i against x_i. Do the data show any trend?

(b) A possible specification for $\mathrm{E}(Y)$ is

$$\mathrm{E}(Y_i) = \exp(\beta_1 + \beta_2 x_i)$$

which will ensure that $\mathrm{E}(Y)$ is non-negative for all values of the parameters and all values of x. Which link function is appropriate in this case?

(c) The exponential distribution is often used to describe survival times. The probability distribution is $f(y;\theta) = \theta e^{-y\theta}$. This is a special case of the gamma distribution with shape parameter $\phi = 1$. Show that $\mathrm{E}(Y) = \theta$ and $\mathrm{var}(Y) = \theta^2$. Fit a model with the equation for $\mathrm{E}(Y_i)$ given in (b) and the exponential distribution using appropriate statistical software.

(d) For the model fitted in (c), compare the observed values y_i and fitted values $\widehat{y}_i = \exp(\widehat{\beta}_1 + \widehat{\beta}_2 x_i)$ and use the standardized residuals $r_i = (y_i - \widehat{y}_i)/\widehat{y}_i$ to investigate the adequacy of the model. (Note: $\widehat{y}_i$ is used as the denominator of r_i because it is an estimate of the standard deviation of Y_i – see (c) above.)

4.3 Let $Y_1, ..., Y_N$ be a random sample from the Normal distribution $Y_i \sim N(\log\beta, \sigma^2)$ where σ^2 is known. Find the maximum likelihood estimator of β from first principles. Also verify equations (4.18) and (4.25) in this case.

5

Inference

5.1 Introduction

The two main tools of statistical inference are confidence intervals and hypothesis tests. Their derivation and use for generalized linear models are covered in this chapter.

Confidence intervals, also known as **interval estimates**, are increasingly regarded as more useful than hypothesis tests because the width of a confidence interval provides a measure of the precision with which inferences can be made. It does so in a way which is conceptually simpler than the power of a statistical test (Altman et al., 2000).

Hypothesis tests in a statistical modelling framework are performed by comparing how well two related models fit the data (see the examples in Chapter 2). For generalized linear models, the two models should have the same probability distribution and the same link function but the linear component of one model has more parameters than the other. The simpler model, corresponding to the null hypothesis H_0, must be a special case of the other more general model. If the simpler model fits the data as well as the more general model does, then it is preferred on the grounds of parsimony and H_0 is retained. If the more general model fits significantly better, then H_0 is rejected in favor of an alternative hypothesis H_1 which corresponds to the more general model. To make these comparisons, we use summary statistics to describe how well the models fit the data. These **goodness of fit statistics** may be based on the maximum value of the likelihood function, the maximum value of the log-likelihood function, the minimum value of the sum of squares criterion or a composite statistic based on the residuals. The process and logic can be summarized as follows:

1. Specify a model M_0 corresponding to H_0. Specify a more general model M_1 (with M_0 as a special case of M_1).
2. Fit M_0 and calculate the goodness of fit statistic G_0. Fit M_1 and calculate the goodness of fit statistic G_1.
3. Calculate the improvement in fit, usually $G_1 - G_0$ but G_1/G_0 is another possibility.
4. Use the sampling distribution of $G_1 - G_0$ (or some related statistic) to test the null hypothesis that $G_1 = G_0$ against the alternative hypothesis $G_1 \neq G_0$.
5. If the hypothesis that $G_1 = G_0$ is not rejected, then H_0 is not rejected and M_0 is the preferred model. If the hypothesis $G_1 = G_0$ is rejected then H_0 is rejected and M_1 is regarded as the better model.

For both forms of inference, sampling distributions are required. To calculate a confidence interval, the sampling distribution of the estimator is required. To test a hypothesis, the sampling distribution of the goodness of fit statistic is required. This chapter is about the relevant sampling distributions for generalized linear models.

If the response variables are Normally distributed, the sampling distributions used for inference can often be determined exactly. For other distributions we need to rely on large-sample asymptotic results based on the Central Limit Theorem. The rigorous development of these results requires careful attention to various regularity conditions. For independent observations from distributions which belong to the exponential family, and in particular for generalized linear models, the necessary conditions are indeed satisfied. In this book we consider only the major steps and not the finer points involved in deriving the sampling distributions. Details of the distribution theory for generalized linear models are given by Fahrmeir and Kaufman (1985).

The basic idea is that under appropriate conditions, if S is a statistic of interest, then approximately

$$\frac{S - \mathrm{E}(S)}{\sqrt{\mathrm{var}(S)}} \sim N(0,1)$$

or equivalently

$$\frac{[S - \mathrm{E}(S)]^2}{\mathrm{var}(S)} \sim \chi^2(1)$$

where $\mathrm{E}(S)$ and $\mathrm{var}(S)$ are the expectation and variance of S respectively.

If there is a vector of statistics of interest $\mathbf{s} = \begin{bmatrix} S_1 \\ \vdots \\ S_p \end{bmatrix}$ with asymptotic expectation $\mathrm{E}(\mathbf{s})$ and asymptotic variance-covariance matrix $\mathbf{V}$, then approximately

$$[\mathbf{s} - \mathrm{E}(\mathbf{s})]^T \mathbf{V}^{-1} [\mathbf{s} - \mathrm{E}(\mathbf{s})] \sim \chi^2(p) \qquad (5.1)$$

provided $\mathbf{V}$ is non-singular so a unique inverse matrix $\mathbf{V}^{-1}$exists.

5.2 Sampling distribution for score statistics

Suppose $Y_1, ..., Y_N$ are independent random variables in a generalized linear model with parameters $\boldsymbol{\beta}$ where $\mathrm{E}(Y_i) = \mu_i$ and $g(\mu_i) = \mathbf{x}_i^T \boldsymbol{\beta} = \eta_i$. From equation (4.18) the score statistics are

$$U_j = \frac{\partial l}{\partial \beta_j} = \sum_{i=1}^{N} \left[\frac{(Y_i - \mu_i)}{\mathrm{var}(Y_i)} x_{ij} \left(\frac{\partial \mu_i}{\partial \eta_i} \right) \right] \qquad \text{for } j = 1, ..., p.$$

As $\mathrm{E}(Y_i) = \mu_i$ for all i,

$$\mathrm{E}(U_j) = 0 \qquad \text{for } j = 1, ..., p \qquad (5.2)$$

consistent with the general result (3.14). The variance-covariance matrix of the score statistics is the information matrix $\mathfrak{I}$ with elements

$$\mathfrak{I}_{jk} = \mathrm{E}[U_j U_k]$$

given by equation (4.20).

If there is only one parameter β, the score statistic has the asymptotic sampling distribution

$$\frac{U}{\sqrt{\mathfrak{I}}} \sim N(0,1), \text{ or equivalently } \frac{U^2}{\mathfrak{I}} \sim \chi^2(1)$$

because $\mathrm{E}(U) = 0$ and $\mathrm{var}(U) = \mathfrak{I}$.

If there is a vector of parameters

$$\boldsymbol{\beta} = \begin{bmatrix} \beta_1 \\ \vdots \\ \beta_p \end{bmatrix} \text{ then the score vector } \mathbf{U} = \begin{bmatrix} U_1 \\ \vdots \\ U_p \end{bmatrix}$$

has the multivariate Normal distribution $\mathbf{U} \sim \mathbf{N}(\mathbf{0}, \mathfrak{I})$, at least asymptotically, and so

$$\mathbf{U}^T \mathfrak{I}^{-1} \mathbf{U} \sim \chi^2(p) \tag{5.3}$$

for large samples.

5.2.1 Example: Score statistic for the Normal distribution

Let $Y_1, ..., Y_N$ be independent, identically distributed random variables with $Y_i \sim N(\mu, \sigma^2)$ where σ^2 is a known constant. The log-likelihood function is

$$l = -\frac{1}{2\sigma^2}\sum_{i=1}^{N}(y_i - \mu)^2 - N\log(\sigma\sqrt{2\pi}).$$

The score statistic is

$$U = \frac{dl}{d\mu} = \frac{1}{\sigma^2}\sum(Y_i - \mu) = \frac{N}{\sigma^2}(\overline{Y} - \mu)$$

so the maximum likelihood estimator, obtained by solving the equation $U = 0$, is $\widehat{\mu} = \overline{Y}$. The expected value of the statistic U is

$$\mathrm{E}(U) = \frac{1}{\sigma^2}\sum[\mathrm{E}(Y_i) - \mu]$$

from equation (1.2). As $\mathrm{E}(Y_i) = \mu$, it follows that $\mathrm{E}(U) = 0$ as expected. The variance of U is

$$\mathfrak{I} = \mathrm{var}(U) = \frac{1}{\sigma^4}\sum \mathrm{var}(Y_i) = \frac{N}{\sigma^2}$$

from equation (1.3) and $\mathrm{var}(Y_i) = \sigma^2$. Therefore

$$\frac{U}{\sqrt{\mathfrak{I}}} = \frac{(\overline{Y} - \mu)}{\sigma/\sqrt{N}}.$$

According to result (5.1) this has the asymptotic distribution $N(0,1)$. In fact, the result is exact because $\overline{Y} \sim N(\mu, \sigma^2/N)$ (see Exercise 1.4(a)). Similarly

$$U^T \mathfrak{J}^{-1} U = \frac{U^2}{\mathfrak{J}} = \frac{(Y-\mu)^2}{\sigma^2/N} \sim \chi^2(1)$$

is an exact result.

The sampling distribution of U can be used make inferences about μ. For example, a 95% confidence interval for μ is $\overline{y} \pm 1.96\sigma/\sqrt{N}$, where σ is assumed to be known.

5.2.2 *Example: Score statistic for the binomial distribution*

If $Y \sim binomial(n, \pi)$ the log-likelihood function is

$$l(\pi; y) = y \log \pi + (n-y)\log(1-\pi) + \log \binom{n}{y}$$

so the score statistic is

$$U = \frac{dl}{d\pi} = \frac{Y}{\pi} - \frac{n-Y}{1-\pi} = \frac{Y - n\pi}{\pi(1-\pi)}.$$

But $\mathrm{E}(Y) = n\pi$ and so $\mathrm{E}(U) = 0$ as expected. Also $\mathrm{var}(Y) = n\pi(1-\pi)$ so

$$\mathfrak{J} = \mathrm{var}(U) = \frac{1}{\pi^2(1-\pi)^2}\mathrm{var}(Y) = \frac{n}{\pi(1-\pi)}$$

and hence

$$\frac{U}{\sqrt{\mathfrak{J}}} = \frac{Y - n\pi}{\sqrt{n\pi(1-\pi)}} \sim N(0,1)$$

approximately. This is the Normal approximation to binomial distribution (without any continuity correction). It is used to find confidence intervals for, and test hypotheses about, π.

5.3 Taylor series approximations

To obtain the asymptotic sampling distributions for various other statistics it is useful to use Taylor series approximations. The Taylor series approximation for a function $f(x)$ of a single variable x about a value t is

$$f(x) = f(t) + (x-t)\left[\frac{df}{dx}\right]_{x=t} + \frac{1}{2}(x-t)^2\left[\frac{d^2 f}{dx^2}\right]_{x=t} + \ldots$$

provided that x is near t.

For a log-likelihood function of a single parameter β the first three terms of the Taylor series approximation near an estimate b are

$$l(\beta) = l(b) + (\beta - b)U(b) + \frac{1}{2}(\beta - b)^2 U'(b)$$

where $U(b) = dl/d\beta$ is the score function evaluated at $\beta = b$. If $U' = d^2l/d\beta^2$ is approximated by its expected value $\mathrm{E}(U') = -\Im$, the approximation becomes

$$l(\beta) = l(b) + (\beta - b)U(b) - \frac{1}{2}(\beta - b)^2\Im(b)$$

where $\Im(b)$ is the information evaluated at $\beta = b$. The corresponding approximation for the log-likelihood function for a vector parameter $\boldsymbol{\beta}$ is

$$l(\boldsymbol{\beta}) = l(\mathbf{b}) + (\boldsymbol{\beta} - \mathbf{b})^T\mathbf{U}(\mathbf{b}) - \frac{1}{2}(\boldsymbol{\beta} - \mathbf{b})^T\Im(\mathbf{b})(\boldsymbol{\beta} - \mathbf{b}) \tag{5.4}$$

where $\mathbf{U}$ is the vector of scores and $\Im$ is the information matrix.

For the score function of a single parameter β the first two terms of the Taylor series approximation near an estimate b give

$$U(\beta) = U(b) + (\beta - b)U'(b).$$

If U' is approximated by $\mathrm{E}(U') = -\Im$ we obtain

$$U(\beta) = U(b) - (\beta - b)\Im(b).$$

The corresponding expression for a vector parameter $\boldsymbol{\beta}$ is

$$\mathbf{U}(\boldsymbol{\beta}) = \mathbf{U}(\mathbf{b}) - \Im(\mathbf{b})(\boldsymbol{\beta} - \mathbf{b}). \tag{5.5}$$

5.4 Sampling distribution for maximum likelihood estimators

Equation (5.5) can be used to obtain the sampling distribution of the maximum likelihood estimator $\mathbf{b} = \widehat{\boldsymbol{\beta}}$. By definition, $\mathbf{b}$ is the estimator which maximizes $l(\mathbf{b})$ and so $\mathbf{U}(\mathbf{b}) = \mathbf{0}$. Therefore

$$\mathbf{U}(\boldsymbol{\beta}) = -\Im(\mathbf{b})(\boldsymbol{\beta} - \mathbf{b})$$

or equivalently,

$$(\mathbf{b} - \boldsymbol{\beta}) = \Im^{-1}\mathbf{U}$$

provided that $\Im$ is non-singular. If $\Im$ is regarded as constant then $\mathrm{E}(\mathbf{b} - \boldsymbol{\beta}) = \mathbf{0}$ because $\mathrm{E}(\mathbf{U}) = \mathbf{0}$ by equation (5.2). Therefore $\mathrm{E}(\mathbf{b}) = \boldsymbol{\beta}$, at least asymptotically, so $\mathbf{b}$ is a consistent estimator of $\boldsymbol{\beta}$. The variance-covariance matrix for $\mathbf{b}$ is

$$\mathrm{E}\left[(\mathbf{b} - \boldsymbol{\beta})(\mathbf{b} - \boldsymbol{\beta})^T\right] = \Im^{-1}\mathrm{E}(\mathbf{U}\mathbf{U}^T)\Im = \Im^{-1} \tag{5.6}$$

because $\Im = \mathrm{E}(\mathbf{U}\mathbf{U}^T)$ and $(\Im^{-1})^T = \Im^{-1}$ as $\Im$ is symmetric.

The asymptotic sampling distribution for $\mathbf{b}$, by (5.1), is

$$(\mathbf{b} - \boldsymbol{\beta})^T\Im(\mathbf{b})(\mathbf{b} - \boldsymbol{\beta}) \sim \chi^2(p). \tag{5.7}$$

This is the **Wald statistic**. For the one-parameter case, the more commonly used form is

$$b \sim N(\beta, \Im^{-1}). \tag{5.8}$$

If the response variables in the generalized linear model are Normally distributed then (5.7) and (5.8) are exact results (see Example 5.4.1 below).

5.4.1 Example: Maximum likelihood estimators for the Normal linear model

Consider the model

$$\mathrm{E}(Y_i) = \mu_i = \mathbf{x}_i^T\boldsymbol{\beta} \quad ; \quad Y_i \sim N(\mu_i, \sigma^2) \tag{5.9}$$

where the Y_i's are N independent random variables and $\boldsymbol{\beta}$ is a vector of p parameters ($p < N$). This is a generalized linear model with the identity function as the link function. This model is discussed in more detail in Chapter 6.

As the link function is the identity, in equation (4.16) $\mu_i = \eta_i$ and so $\partial\mu_i/\partial\eta_i = 0$. The elements of the information matrix, given in equation (4.20), have the simpler form

$$\mathfrak{I}_{jk} = \sum_{i=1}^{N} \frac{x_{ij}x_{ik}}{\sigma^2}$$

because $\mathrm{var}(Y_i) = \sigma^2$. Therefore the information matrix can be written as

$$\mathfrak{I} = \frac{1}{\sigma^2}\mathbf{X}^T\mathbf{X}. \tag{5.10}$$

Similarly the expression in (4.24) has the simpler form

$$z_i = \sum_{k=1}^{p} x_{ik}b_k^{(m-1)} + (y_i - \mu_i).$$

But μ_i evaluated at $\mathbf{b}^{(m-1)}$ is $\mathbf{x}_i^T\mathbf{b}^{(m-1)} = \sum_{k=1}^{p} x_{ik}b_k^{(m-1)}$. Therefore $z_i = y_i$ in this case.

The estimating equation (4.25) is

$$\frac{1}{\sigma^2}\mathbf{X}^T\mathbf{X}\mathbf{b} = \frac{1}{\sigma^2}\mathbf{X}^T\mathbf{y}$$

and hence the maximum likelihood estimator is

$$\mathbf{b} = (\mathbf{X}^T\mathbf{X})^{-1}\mathbf{X}^T\mathbf{y}. \tag{5.11}$$

The model (5.9) can be written in vector notation as $\mathbf{y} \sim \mathbf{N}(\mathbf{X}\boldsymbol{\beta}, \sigma^2\mathbf{I})$ where $\mathbf{I}$ is the $N \times N$ unit matrix with ones on the diagonal and zeros elsewhere. From (5.11)

$$\mathrm{E}(\mathbf{b}) = (\mathbf{X}^T\mathbf{X})^{-1}(\mathbf{X}^T\mathbf{X}\boldsymbol{\beta}) = \boldsymbol{\beta}$$

so $\mathbf{b}$ is an unbiased estimator of $\boldsymbol{\beta}$.

To obtain the variance-covariance matrix for $\mathbf{b}$ we use

$$\begin{aligned}\mathbf{b} - \boldsymbol{\beta} &= (\mathbf{X}^T\mathbf{X})^{-1}\mathbf{X}^T\mathbf{y} - \boldsymbol{\beta} \\ &= (\mathbf{X}^T\mathbf{X})^{-1}\mathbf{X}^T(\mathbf{y} - \mathbf{X}\boldsymbol{\beta}).\end{aligned}$$

Hence

$$\begin{aligned} \mathrm{E}\left[(\mathbf{b}-\boldsymbol{\beta})(\mathbf{b}-\boldsymbol{\beta})^T\right] &= (\mathbf{X}^T\mathbf{X})^{-1}\mathbf{X}^T\mathrm{E}\left[(\mathbf{y}-\mathbf{X}\boldsymbol{\beta})(\mathbf{y}-\mathbf{X}\boldsymbol{\beta})^T\right]\mathbf{X}(\mathbf{X}^T\mathbf{X})^{-1} \\ &= (\mathbf{X}^T\mathbf{X})^{-1}\mathbf{X}^T\left[\mathrm{var}(\mathbf{y})\right]\mathbf{X}(\mathbf{X}^T\mathbf{X})^{-1} \\ &= \sigma^2(\mathbf{X}^T\mathbf{X})^{-1} \end{aligned}$$

But $\sigma^2(\mathbf{X}^T\mathbf{X})^{-1} = \mathfrak{I}^{-1}$ from (5.10) so the variance-covariance matrix for **b** is $\mathfrak{I}^{-1}$ as in (5.6).

The maximum likelihood estimator **b** is a linear combination of the elements Y_i of $\mathbf{y}$, from (5.11). As the Y_is are Normally distributed, from the results in Section 1.4.1, the elements of **b** are also Normally distributed. Hence the exact sampling distribution of **b**, in this case, is

$$\mathbf{b} \sim N(\boldsymbol{\beta}, \mathfrak{I}^{-1})$$

or

$$(\mathbf{b}-\boldsymbol{\beta})^T\mathfrak{I}(\mathbf{b}-\boldsymbol{\beta}) \sim \chi^2(p).$$

5.5 Log-likelihood ratio statistic

One way of assessing the adequacy of a model is to compare it with a more general model with the maximum number of parameters that can be estimated. This is called a **saturated model**. It is a generalized linear model with the same distribution and same link function as the model of interest.

If there are N observations $Y_i, i = 1, \ldots, N$, all with potentially different values for the linear component $\mathbf{x}_i^T\boldsymbol{\beta}$, then a saturated model can be specified with N parameters. This is also called a **maximal** or **full model**.

If some of the observations have the same linear component or covariate pattern, i.e., they correspond to the same combination of factor levels and have the same values of any continuous explanatory variables, they are called **replicates**. In this case, the maximum number of parameters that can be estimated for the saturated model is equal to the number of potentially different linear components, which may be less than N.

In general, let m denote the maximum number of parameters that can be estimated. Let $\boldsymbol{\beta}_{\max}$ denote the parameter vector for the saturated model and $\mathbf{b}_{\max}$ denote the maximum likelihood estimator of $\boldsymbol{\beta}_{\max}$. The likelihood function for the saturated model evaluated at $\mathbf{b}_{\max}$, $L(\mathbf{b}_{\max};\mathbf{y})$, will be larger than any other likelihood function for these observations, with the same assumed distribution and link function, because it provides the most complete description of the data. Let $L(\mathbf{b};\mathbf{y})$ denote the maximum value of the likelihood function for the model of interest. Then the likelihood ratio

$$\lambda = \frac{L(\mathbf{b}_{\max};\mathbf{y})}{L(\mathbf{b};\mathbf{y})}$$

provides a way of assessing the goodness of fit for the model. In practice, the logarithm of the likelihood ratio, which is the difference between the log-likelihood functions,

$$\log \lambda = l(\mathbf{b}_{\max}; \mathbf{y}) - l(\mathbf{b}; \mathbf{y})$$

is used. Large values of $\log \lambda$ suggest that the model of interest is a poor description of the data relative to the saturated model. To determine the critical region for $\log \lambda$ we need its sampling distribution.

In the next section we see that $2 \log \lambda$ has a chi-squared distribution. Therefore $2 \log \lambda$ rather than $\log \lambda$ is the more commonly used statistic. It was called the **deviance** by Nelder and Wedderburn (1972).

5.6 Sampling distribution for the deviance

The deviance, also called the **log likelihood (ratio) statistic**, is

$$D = 2[l(\mathbf{b}_{\max}; \mathbf{y}) - l(\mathbf{b}; \mathbf{y})].$$

From equation (5.4), if $\mathbf{b}$ is the maximum likelihood estimator of the parameter $\boldsymbol{\beta}$ (so that $\mathbf{U}(\mathbf{b}) = \mathbf{0}$)

$$l(\boldsymbol{\beta}) - l(\mathbf{b}) = -\frac{1}{2}(\boldsymbol{\beta} - \mathbf{b})^T \mathfrak{I}(\mathbf{b})(\boldsymbol{\beta} - \mathbf{b})$$

approximately. Therefore the statistic

$$2[l(\mathbf{b}; \mathbf{y}) - l(\boldsymbol{\beta}; \mathbf{y})] = (\boldsymbol{\beta} - \mathbf{b})^T \mathfrak{I}(\mathbf{b})(\boldsymbol{\beta} - \mathbf{b}),$$

which has the chi-squared distribution $\chi^2(p)$ where p is the number of parameters, from (5.7).

From this result the sampling distribution for the deviance can be derived:

$$\begin{aligned} D &= 2[l(\mathbf{b}_{\max}; \mathbf{y}) - l(\mathbf{b}; \mathbf{y})] \\ &= 2[l(\mathbf{b}_{\max}; \mathbf{y}) - l(\boldsymbol{\beta}_{\max}; \mathbf{y})] \\ &\quad -2[l(\mathbf{b}; \mathbf{y}) - l(\boldsymbol{\beta}; \mathbf{y})] + 2[l(\boldsymbol{\beta}_{\max}; \mathbf{y}) - l(\boldsymbol{\beta}; \mathbf{y})]. \qquad (5.12) \end{aligned}$$

The first term in square brackets in (5.12) has the distribution $\chi^2(m)$ where m is the number of parameters in the saturated model. The second term has the distribution $\chi^2(p)$ where p is the number of parameters in the model of interest. The third term, $\upsilon = 2[l(\boldsymbol{\beta}_{\max}; \mathbf{y}) - l(\boldsymbol{\beta}; \mathbf{y})]$, is a positive constant which will be near zero if the model of interest fits the data almost as well as the saturated model fits. Therefore the sampling distribution of the deviance is, approximately,

$$D \sim \chi^2(m - p, \upsilon)$$

where υ is the non-centrality parameter, by the results in Section 1.5. The deviance forms the basis for most hypothesis testing for generalized linear models. This is described in Section 5.7.

If the response variables Y_i are Normally distributed then D has a chi-squared distribution exactly. In this case, however, D depends on $\text{var}(Y_i) = \sigma^2$ which, in practice, is usually unknown. This means that D cannot be used directly as a goodness of fit statistic (see Example 5.6.2).

For Y_i's with other distributions, the sampling distribution of D may be only approximately chi-squared. However for the binomial and Poisson distributions, for example, D can be calculated and used directly as a goodness of fit statistic (see Example 5.6.1 and 5.6.3).

5.6.1 *Example: Deviance for a binomial model*

If the response variables $Y_1, ..., Y_N$ are independent and $Y_i \sim binomial(n_i, \pi_i)$, then the log-likelihood function is

$$l(\boldsymbol{\beta}; \mathbf{y}) = \sum_{i=1}^{N} \left[y_i \log \pi_i - y_i \log(1 - \pi_i) + n_i \log(1 - \pi_i) + \log \binom{n_i}{y_i} \right].$$

For a saturated model, the π_i's are all different so $\boldsymbol{\beta} = [\pi_1, ..., \pi_N]^T$. The maximum likelihood estimates are $\widehat{\pi}_i = y_i/n_i$ so the maximum value of the log-likelihood function is

$$l(\mathbf{b}_{\max}; \mathbf{y}) = \sum \left[y_i \log \left(\frac{y_i}{n_i} \right) - y_i \log(\frac{n_i - y_i}{n_i}) + n_i \log(\frac{n_i - y_i}{n_i}) + \log \binom{n_i}{y_i} \right].$$

For any other model with $p < N$ parameters, let $\widehat{\pi}_i$ denote the maximum likelihood estimates for the probabilities and let $\widehat{y}_i = n_i \widehat{\pi}_i$ denote the fitted values. Then the log-likelihood function evaluated at these values is

$$l(\mathbf{b}; \mathbf{y}) = \sum \left[y_i \log \left(\frac{\widehat{y}_i}{n_i} \right) - y_i \log(\frac{n_i - \widehat{y}_i}{n_i}) + n_i \log(\frac{n_i - \widehat{y}_i}{n_i}) + \log \binom{n_i}{y_i} \right].$$

Therefore the deviance is

$$\begin{aligned} D &= 2\left[l(\mathbf{b}_{\max}; \mathbf{y}) - l(\mathbf{b}; \mathbf{y})\right] \\ &= 2\sum_{i=1}^{N} \left[y_i \log \left(\frac{y_i}{\widehat{y}_i} \right) + (n_i - y_i) \log(\frac{n_i - y_i}{n_i - \widehat{y}_i}) \right]. \end{aligned}$$

5.6.2 *Example: Deviance for a Normal linear model*

Consider the model

$$\mathrm{E}(Y_i) = \mu_i = \mathbf{x}_i^T \boldsymbol{\beta} \quad ; \quad Y_i \sim N(\mu_i, \sigma^2), i = 1, ..., N$$

where the Y_i's are independent. The log-likelihood function is

$$l(\boldsymbol{\beta}; \mathbf{y}) = -\frac{1}{2\sigma^2} \sum_{i=1}^{N} (y_i - \mu_i)^2 - \frac{1}{2} N \log(2\pi\sigma^2).$$

For a saturated model all the μ_i's can be different so $\boldsymbol{\beta}$ has N elements $\mu_1, ..., \mu_N$. By differentiating the log-likelihood function with respect to each μ_i and solving the estimating equations, we obtain $\widehat{\mu}_i = y_i$. Therefore the maximum value of the log-likelihood function for the saturated model is

$$l(\mathbf{b}_{\max}; \mathbf{y}) = -\frac{1}{2} N \log(2\pi\sigma^2).$$

For any other model with $p < N$ parameters, let

$$\mathbf{b} = (\mathbf{X}^T\mathbf{X})^{-1}\mathbf{X}^T\mathbf{y}$$

be the maximum likelihood estimator (from equation 5.11). The corresponding maximum value for the log-likelihood function is

$$l(\mathbf{b};\mathbf{y}) = -\frac{1}{2\sigma^2}\sum\left(y_i - \mathbf{x}_i^T\mathbf{b}\right)^2 - \frac{1}{2}N\log(2\pi\sigma^2).$$

Therefore the deviance is

$$\begin{aligned} D &= 2[l(\mathbf{b}_{\max};\mathbf{y}) - l(\mathbf{b};\mathbf{y})] \\ &= \frac{1}{\sigma^2}\sum_{i=1}^{N}(y_i - \mathbf{x}_i^T\mathbf{b})^2 && (5.13) \\ &= \frac{1}{\sigma^2}\sum_{i=1}^{N}(y_i - \widehat{\mu}_i)^2 && (5.14) \end{aligned}$$

where $\widehat{\mu}_i$ denotes the fitted value $\mathbf{x}_i^T\mathbf{b}$.

In the particular case where there is only one parameter, for example when $\mathrm{E}(Y_i) = \mu$ for all i, $\mathbf{X}$ is a vector of N ones and so $b = \widehat{\mu} = \sum_{i=1}^{N} y_i/N = \overline{y}$ and $\widehat{\mu}_i = \overline{y}$ for all i. Therefore

$$D = \frac{1}{\sigma^2}\sum_{i=1}^{N}(y_i - \overline{y})^2.$$

But this statistic is related to the sample variance S^2

$$S^2 = \frac{1}{N-1}\sum_{i=1}^{N}(y_i - \overline{y})^2 = \frac{\sigma^2 D}{N-1}.$$

From Exercise 1.4(d) $(N-1)S^2/\sigma^2 \sim \chi^2(N-1)$ so $D \sim \chi^2(N-1)$ exactly.

More generally, from (5.13)

$$\begin{aligned} D &= \frac{1}{\sigma^2}\sum(y_i - \mathbf{x}_i^T\mathbf{b})^2 \\ &= \frac{1}{\sigma^2}(\mathbf{y} - \mathbf{X}\mathbf{b})^T(\mathbf{y} - \mathbf{X}\mathbf{b}) \end{aligned}$$

where the design matrix $\mathbf{X}$ has rows $\mathbf{x}_i$. The term $(\mathbf{y} - \mathbf{X}\mathbf{b})$ can be written as

$$\begin{aligned} \mathbf{y} - \mathbf{X}\mathbf{b} &= \mathbf{y} - \mathbf{X}(\mathbf{X}^T\mathbf{X})^{-1}\mathbf{X}^T\mathbf{y} \\ &= [\mathbf{I} - \mathbf{X}(\mathbf{X}^T\mathbf{X})^{-1}\mathbf{X}^T]\mathbf{y} = [\mathbf{I} - \mathbf{H}]\mathbf{y} \end{aligned}$$

where $\mathbf{H} = \mathbf{X}(\mathbf{X}^T\mathbf{X})^{-1}\mathbf{X}^T$, which is called the **'hat' matrix**. Therefore the quadratic form in D can be written as

$$(\mathbf{y} - \mathbf{X}\mathbf{b})^T(\mathbf{y} - \mathbf{X}\mathbf{b}) = \{[\mathbf{I} - \mathbf{H}]\mathbf{y}\}^T[\mathbf{I} - \mathbf{H}]\mathbf{y} = \mathbf{y}^T[\mathbf{I} - \mathbf{H}]\mathbf{y}$$

because $\mathbf{H}$ is idempotent (i.e., $\mathbf{H} = \mathbf{H}^T$ and $\mathbf{H}\mathbf{H} = \mathbf{H}$). The rank of $\mathbf{I}$ is n and the rank of $\mathbf{H}$ is p so the rank of $\mathbf{I} - \mathbf{H}$ is $n-p$ so, from Section 1.4.2, part

8, D has a chi-squared distribution with $n-p$ degrees of freedom and non-centrality parameter $\lambda = (\mathbf{X}\boldsymbol{\beta})^T(\mathbf{I}-\mathbf{H})(\mathbf{X}\boldsymbol{\beta})/\sigma^2$. But $(\mathbf{I}-\mathbf{H})\mathbf{X} = \mathbf{0}$ so D has the central distribution $\chi^2(N-p)$ exactly (for more details, see Graybill, 1976).

The term **scaled deviance** is sometimes used for

$$\sigma^2 D = \sum (y_i - \widehat{\mu}_i)^2.$$

If the model fits the data well, then $D \sim \chi^2(N-p)$. The expected value for a random variable with the distribution $\chi^2(N-p)$ is $N-p$ (from Section 1.4.2 part 2), so the expected value of D is $N-p$.

This provides an estimate of σ^2 as

$$\widetilde{\sigma}^2 = \frac{\sum (y_i - \widehat{\mu}_i)^2}{N-p}.$$

Some statistical programs, such as Glim, output the scaled deviance for a Normal linear model and call $\widetilde{\sigma}^2$ the scale parameter.

The deviance is also related to the sum of squares of the standardized residuals (see Section 2.3.4)

$$\sum_{i=1}^{N} r_i^2 = \frac{1}{\widehat{\sigma}^2} \sum_{i=1}^{N} (y_i - \widehat{\mu}_i)^2$$

where $\widehat{\sigma}^2$ is an estimate of σ^2. This provides a rough rule of thumb for the overall magnitude of the standardized residuals. If the model fits well so that $D \sim \chi^2(N-p)$, you could expect $\sum r_i^2 = N-p$, approximately.

5.6.3 Example: Deviance for a Poisson model

If the response variables $Y_1, ..., Y_N$ are independent and $Y_i \sim Poisson(\lambda_i)$, the log-likelihood function is

$$l(\boldsymbol{\beta}; \mathbf{y}) = \sum y_i \log \lambda_i - \sum \lambda_i - \sum \log y_i!.$$

For the saturated model, the λ_i's are all different so $\boldsymbol{\beta} = [\lambda_1, ..., \lambda_N]^T$. The maximum likelihood estimates are $\widehat{\lambda}_i = y_i$ and so the maximum value of the log-likelihood function is

$$l(\mathbf{b}_{\max}; \mathbf{y}) = \sum y_i \log y_i - \sum y_i - \sum \log y_i!.$$

Suppose the model of interest has $p < N$ parameters. The maximum likelihood estimator $\mathbf{b}$ can be used to calculate estimates $\widehat{\lambda}_i$ and hence fitted values $\widehat{y}_i = \widehat{\lambda}_i$; because $\mathrm{E}(Y_i) = \lambda_i$. The maximum value of the log-likelihood in this case is

$$l(\mathbf{b}; \mathbf{y}) = \sum y_i \log \widehat{y}_i - \sum \widehat{y}_i - \sum \log y_i!.$$

Therefore the deviance is

$$\begin{aligned} D &= 2[l(\mathbf{b}_{\max}; \mathbf{y}) - l(\mathbf{b}; \mathbf{y})] \\ &= 2\left[\sum y_i \log (y_i/\widehat{y}_i) - \sum (y_i - \widehat{y}_i)\right]. \end{aligned}$$

For most models it can shown that $\sum y_i = \sum \widehat{y}_i$ – see Exercise 9.1. Therefore D can be written in the form

$$D = 2\sum o_i \log(o_i/e_i)$$

if o_i is used to denote the observed value y_i and e_i is used to denote the estimated expected value $\widehat{y}_i$.

The value of D can be calculated from the data in this case (unlike the case for the Normal distribution where D depends on the unknown constant σ^2). This value can be compared with the distribution $\chi^2(N-p)$. The following example illustrates the idea.

The data in Table 5.1 relate to Example 4.4 where a straight line was fitted to Poisson responses. The fitted values are

$$\widehat{y}_i = b_1 + b_2 x_i$$

where $b_1 = 7.45163$ and $b_2 = 4.93530$ (from Table 4.4). The value of D is $D = 2 \times (0.94735 - 0) = 1.8947$ which is small relative to the degrees of freedom, $N - p = 9 - 2 = 7$. In fact, D is below the lower 5% tail of the distribution $\chi^2(7)$ indicating that the model fits the data well – perhaps not surprisingly for such a small set of artificial data!

Table 5.1 *Results from the Poisson regression Example 4.4.*

x_i	y_i	$\widehat{y}_i$	$y_i \log(y_i/\widehat{y}_i)$
-1	2	2.51633	-0.45931
-1	3	2.51633	0.52743
0	6	7.45163	-1.30004
0	7	7.45163	-0.43766
0	8	7.45163	0.56807
0	9	7.45163	1.69913
1	10	12.38693	-2.14057
1	12	12.38693	-0.38082
1	15	12.38693	2.87112
Total	72	72	0.94735

5.7 Hypothesis testing

Hypotheses about a parameter vector $\boldsymbol{\beta}$ of length p can be tested using the sampling distribution of the Wald statistic $(\widehat{\boldsymbol{\beta}} - \boldsymbol{\beta})^T \mathfrak{I}(\widehat{\boldsymbol{\beta}} - \boldsymbol{\beta}) \sim \chi^2(p)$ (from 5.7). Occasionally the score statistic is used: $\mathbf{U}^T \mathfrak{I}^{-1} \mathbf{U} \sim \chi^2(p)$ from (5.3).

An alternative approach, outlined in Section 5.1 and used in Chapter 2, is to compare the goodness of fit of two models. The models need to be **nested** or **hierarchical**, that is, they have the same probability distribution and the same link function but the linear component of the simpler model M_0 is a special case of the linear component of the more general model M_1.

Consider the null hypothesis

$$H_0 : \boldsymbol{\beta} = \boldsymbol{\beta}_0 = \begin{bmatrix} \beta_1 \\ \vdots \\ \beta_q \end{bmatrix}$$

corresponding to model M_0 and a more general hypothesis

$$H_1 : \boldsymbol{\beta} = \boldsymbol{\beta}_1 = \begin{bmatrix} \beta_1 \\ \vdots \\ \beta_p \end{bmatrix}$$

corresponding to M_1, with $q < p < N$.

We can test H_0 against H_1 using the difference of the deviance statistics

$$\begin{aligned} \triangle D &= D_0 - D_1 = 2[l(\mathbf{b}_{\max}; \mathbf{y}) - l(\mathbf{b}_0; \mathbf{y})] - 2[l(\mathbf{b}_{\max}; \mathbf{y}) - l(\mathbf{b}_1; \mathbf{y})] \\ &= 2[l(\mathbf{b}_1; \mathbf{y}) - l(\mathbf{b}_0; \mathbf{y})]. \end{aligned}$$

If both models describe the data well then $D_0 \sim \chi^2(N-q)$ and $D_1 \sim \chi^2(N-p)$ so that $\triangle D \sim \chi^2(p-q)$, provided that certain independence conditions hold. If the value of $\triangle D$ is consistent with the $\chi^2(p-q)$ distribution we would generally choose the model M_0 corresponding to H_0 because it is simpler.

If the value of $\triangle D$ is in the critical region (i.e., greater than the upper tail $100\times\alpha\%$ point of the $\chi^2(p-q)$ distribution) then we would reject H_0 in favor of H_1 on the grounds that model M_1 provides a significantly better description of the data (even though it too may not fit the data particularly well).

Provided that the deviance can be calculated from the data, $\triangle D$ provides a good method for hypothesis testing. The sampling distribution of $\triangle D$ is usually better approximated by the chi-squared distribution than is the sampling distribution of a single deviance.

For models based on the Normal distribution, or other distributions with nuisance parameters that are not estimated, the deviance may not be fully determined from the data. The following example shows how this problem may be overcome.

5.7.1 Example: Hypothesis testing for a Normal linear model

For the Normal linear model

$$\mathrm{E}(Y_i) = \mu_i = \mathbf{x}_i^T \boldsymbol{\beta} \quad ; \quad Y_i \sim N(\mu_i, \sigma^2)$$

for independent random variables $Y_1, ..., Y_N$, the deviance is

$$D = \frac{1}{\sigma^2} \sum_{i=1}^{N} (y_i - \widehat{\mu}_i)^2,$$

from equation (5.14).

Let $\widehat{\mu}_i(0)$ and $\widehat{\mu}_i(1)$ denote the fitted values for model M_0 (corresponding to null hypothesis H_0) and model M_1(corresponding to the alternative hypothesis H_1) respectively. Then

$$D_0 = \frac{1}{\sigma^2}\sum_{i=1}^{N}\left[y_i - \widehat{\mu}_i(0)\right]^2$$

and

$$D_1 = \frac{1}{\sigma^2}\sum_{i=1}^{N}\left[y_i - \widehat{\mu}_i(1)\right]^2.$$

It is usual to assume that M_1 fits the data well (and so H_1 is correct), so that $D_1 \sim \chi^2(N-p)$. If M_0 is also fits well, then $D_0 \sim \chi^2(N-q)$ and so $\triangle D = D_0 - D_1 \sim \chi^2(p-q)$. If M_0 does not fit well (i.e., H_0 is not correct) then $\triangle D$ will have a non-central χ^2 distribution. To eliminate the term σ^2 we use the ratio

$$\begin{aligned} F &= \frac{D_0 - D_1}{p-q} \Big/ \frac{D_1}{N-p} \\ &= \frac{\left\{\sum\left[y_i - \widehat{\mu}_i(0)\right]^2 - \sum\left[y_i - \widehat{\mu}_i(1)\right]^2\right\}/(p-q)}{\sum\left[y_i - \widehat{\mu}_i(1)\right]^2/(N-p)}. \end{aligned}$$

Thus F can be calculated directly from the fitted values. If H_0 is correct, F will have the central $F(p-q, N-p)$ distribution (at least approximately). If H_0 is not correct, the value of F will be larger than expected from the distribution $F(p-q, N-p)$.

A numerical illustration is provided by the example on birthweights and gestational age in Section 2.2.2. The models are given in (2.6) and (2.7). The minimum values of the sums of squares are related to the deviances by $\widehat{S}_0 = \sigma^2 D_0$ and $\widehat{S}_1 = \sigma^2 D_1$. There are $N = 24$ observations. The simpler model (2.6) has $q = 3$ parameters to be estimated and the more general model (2.7) has $p = 4$ parameters to be estimated. From Table 2.5

$$\begin{aligned} D_0 &= 658770.8/\sigma^2 \quad \text{with } N-q = 21 \text{ degrees of freedom} \\ \text{and } D_1 &= 652424.5/\sigma^2 \quad \text{with } N-p = 20 \text{ degrees of freedom.} \end{aligned}$$

Therefore

$$F = \frac{(658770.8 - 652424.5)/1}{652424.5/20} = 0.19$$

which is certainly not significant compared to the $F(1, 20)$ distribution. So the data are consistent with model (2.6) in which birthweight increases with gestational age at the same rate for boys and girls.

5.8 Exercises

5.1 Consider the single response variable Y with $Y \sim binomial(n, \pi)$.

(a) Find the Wald statistic $(\widehat{\pi} - \pi)^T \Im (\widehat{\pi} - \pi)$ where $\widehat{\pi}$ is the maximum likelihood estimator of π and $\Im$ is the information.

(b) Verify that the Wald statistic is the same as the score statistic $U^T \Im^{-1} U$ in this case (see Example 5.2.2).

(c) Find the deviance

$$2[l(\widehat{\pi}; y) - l(\pi; y)].$$

(d) For large samples, both the Wald/score statistic and the deviance approximately have the $\chi^2(1)$ distribution. For $n = 10$ and $y = 3$ use both statistics to assess the adequacy of the models:

(i) $\pi = 0.1$; (ii) $\pi = 0.3$; (iii) $\pi = 0.5$.

Do the two statistics lead to the same conclusions?

5.2 Consider a random sample $Y_1, ..., Y_N$ with the exponential distribution

$$f(y_i; \theta_i) = \theta_i \exp(-y_i \theta_i).$$

Derive the deviance by comparing the maximal model with different values of θ_i for each Y_i and the model with $\theta_i = \theta$ for all i.

5.3 Suppose $Y_1, ..., Y_N$ are independent identically distributed random variables with the Pareto distribution with parameter θ.

(a) Find the maximum likelihood estimator $\widehat{\theta}$ of θ.

(b) Find the Wald statistic for making inferences about θ (Hint: Use the results from Exercise 3.10).

(c) Use the Wald statistic to obtain an expression for an approximate 95% confidence interval for θ.

(d) Random variables Y with the Pareto distribution with the parameter θ can be generated from random numbers U which are uniformly distributed between 0 and 1 using the relationship $Y = (1/U)^{1/\theta}$ (Evans et al., 2000). Use this relationship to generate a sample of 100 values of Y with $\theta = 2$. From these data calculate an estimate $\widehat{\theta}$. Repeat this process 20 times and also calculate 95% confidence intervals for θ. Compare the average of the estimates $\widehat{\theta}$ with $\theta = 2$. How many of the confidence intervals contain θ?

5.4 For the leukemia survival data in Exercise 4.2:

(a) Use the Wald statistic to obtain an approximate 95% confidence interval for the parameter β_1.

(b) By comparing the deviances for two appropriate models, test the null hypothesis $\beta_2 = 0$ against the alternative hypothesis, $\beta_2 \neq 0$. What can you conclude about the use of the initial white blood cell count as a predictor of survival time?

6

Normal Linear Models

6.1 Introduction

This chapter is about models of the form

$$\mathrm{E}(Y_i) = \mu_i = \mathbf{x}_i^T \boldsymbol{\beta} \quad ; \quad Y_i \sim N(\mu_i, \sigma^2) \tag{6.1}$$

where $Y_1, ..., Y_N$ are independent random variables. The link function is the identity function, i.e., $g(\mu_i) = \mu_i$. This model is usually written as

$$\mathbf{y} = \mathbf{X}\boldsymbol{\beta} + \mathbf{e} \tag{6.2}$$

where

$$\mathbf{y} = \begin{bmatrix} Y_1 \\ \vdots \\ Y_N \end{bmatrix}, \quad \mathbf{X} = \begin{bmatrix} \mathbf{x}_1^T \\ \vdots \\ \mathbf{x}_N^T \end{bmatrix}, \quad \boldsymbol{\beta} = \begin{bmatrix} \beta_1 \\ \vdots \\ \beta_p \end{bmatrix}, \quad \mathbf{e} = \begin{bmatrix} e_1 \\ \vdots \\ e_N \end{bmatrix}$$

and the e_i's are independently, identically distributed random variables with $e_i \sim N(0, \sigma^2)$ for $i = 1, ..., N$. Multiple linear regression, analysis of variance (ANOVA) and analysis of covariance (ANCOVA) are all of this form and together are sometimes called **general linear models.**

The coverage in this book is not detailed, rather the emphasis is on those aspects which are particularly relevant for the model fitting approach to statistical analysis. Many books provide much more detail; for example, see Neter et al. (1996).

The chapter begins with a summary of basic results, mainly derived in previous chapters. Then the main issues are illustrated through four numerical examples.

6.2 Basic results

6.2.1 Maximum likelihood estimation

From Section 5.4.1, the maximum likelihood estimator of $\boldsymbol{\beta}$ is given by

$$\mathbf{b} = (\mathbf{X}^T\mathbf{X})^{-1}\mathbf{X}^T\mathbf{y}. \tag{6.3}$$

provided $(\mathbf{X}^T\mathbf{X})$ is non-singular. As $\mathrm{E}(\mathbf{b}) = \boldsymbol{\beta}$, the estimator is unbiased. It has variance-covariance matrix $\sigma^2(\mathbf{X}^T\mathbf{X})^{-1} = \mathfrak{I}^{-1}$.

In the context of generalized linear models, σ^2 is treated as a nuisance parameter. However it can be shown that

$$\widehat{\sigma}^2 = \frac{1}{N-p}(\mathbf{y} - \mathbf{X}\mathbf{b})^T(\mathbf{y} - \mathbf{X}\mathbf{b}) \tag{6.4}$$

is an unbiased estimator of σ^2 and this can be used to estimate $\mathfrak{I}$ and hence make inferences about $\mathbf{b}$.

6.2.2 *Least squares estimation*

If $\mathrm{E}(\mathbf{y}) = \mathbf{Xb}$ and $\mathrm{E}[(\mathbf{y} - \mathbf{Xb})(\mathbf{y} - \mathbf{Xb})^T] = \mathbf{V}$ where $\mathbf{V}$ is known, we can obtain the least squares estimator $\widetilde{\boldsymbol{\beta}}$ of $\boldsymbol{\beta}$ without making any further assumptions about the distribution of $\mathbf{y}$. We minimize

$$S_w = (\mathbf{y} - \mathbf{Xb})^T \mathbf{V}^{-1} (\mathbf{y} - \mathbf{Xb}).$$

The solution of

$$\frac{\partial S_w}{\partial \boldsymbol{\beta}} = -2\mathbf{X}^T \mathbf{V}^{-1} (\mathbf{y} - \mathbf{Xb}) = \mathbf{0}$$

is

$$\widetilde{\boldsymbol{\beta}} = (\mathbf{X}^T \mathbf{V}^{-1} \mathbf{X})^{-1} \mathbf{X}^T \mathbf{V}^{-1} \mathbf{y},$$

provided the matrix inverses exist. In particular, for model (6.1), where the elements of $\mathbf{y}$ are independent and have a common variance then

$$\widetilde{\boldsymbol{\beta}} = (\mathbf{X}^T \mathbf{X})^{-1} \mathbf{X}^T \mathbf{y}.$$

So in this case, maximum likelihood estimators and least squares estimators are the same.

6.2.3 *Deviance*

From Section 5.6.1

$$\begin{aligned} D &= \frac{1}{\sigma^2} (\mathbf{y} - \mathbf{Xb})^T (\mathbf{y} - \mathbf{Xb}) \\ &= \frac{1}{\sigma^2} (\mathbf{y}^T \mathbf{y} - 2\mathbf{b}^T \mathbf{X}^T \mathbf{y} + \mathbf{b}^T \mathbf{X}^T \mathbf{Xb}) \\ &= \frac{1}{\sigma^2} (\mathbf{y}^T \mathbf{y} - \mathbf{b}^T \mathbf{X}^T \mathbf{y}) \qquad (6.5) \end{aligned}$$

because $\mathbf{X}^T \mathbf{Xb} = \mathbf{X}^T \mathbf{y}$ from equation (6.3).

6.2.4 *Hypothesis testing*

Consider a null hypothesis H_0 and a more general hypothesis H_1 specified as follows

$$H_0 : \boldsymbol{\beta} = \boldsymbol{\beta}_0 = \begin{bmatrix} \beta_1 \\ \vdots \\ \beta_q \end{bmatrix} \quad \text{and} \quad H_1 : \boldsymbol{\beta} = \boldsymbol{\beta}_1 = \begin{bmatrix} \beta_1 \\ \vdots \\ \beta_p \end{bmatrix}$$

where $q < p < N$. Let $\mathbf{X}_0$ and $\mathbf{X}_1$ denote the corresponding design matrices, $\mathbf{b}_0$ and $\mathbf{b}_1$ the maximum likelihood estimators, and D_0 and D_1 the deviances. We test H_0 against H_1 using

$$\begin{aligned} \Delta D &= D_0 - D_1 = \frac{1}{\sigma^2} \left[(\mathbf{y}^T \mathbf{y} - \mathbf{b}_0^T \mathbf{X}_0^T \mathbf{y}) - (\mathbf{y}^T \mathbf{y} - \mathbf{b}_1^T \mathbf{X}_1^T \mathbf{y}) \right] \\ &= \frac{1}{\sigma^2} (\mathbf{b}_1^T \mathbf{X}_1^T \mathbf{y} - \mathbf{b}_0^T \mathbf{X}_0^T \mathbf{y}) \end{aligned}$$

Table 6.1 *Analysis of Variance table.*

Source of variance	Degrees of freedom	Sum of squares	Mean square
Model with β_0	q	$\mathbf{b}_0^T\mathbf{X}_0^T\mathbf{y}$	
Improvement due to model with β_1	$p-q$	$\mathbf{b}_1^T\mathbf{X}_1^T\mathbf{y}-\mathbf{b}_0^T\mathbf{X}_0^T\mathbf{y}$	$\dfrac{\mathbf{b}_1^T\mathbf{X}_1^T\mathbf{y}-\mathbf{b}_0^T\mathbf{X}_0^T\mathbf{y}}{p-q}$
Residual	$N-p$	$\mathbf{y}^T\mathbf{y}-\mathbf{b}_1^T\mathbf{X}_1^T\mathbf{y}$	$\dfrac{\mathbf{y}^T\mathbf{y}-\mathbf{b}_1^T\mathbf{X}_1^T\mathbf{y}}{N-p}$
Total	N	$\mathbf{y}^T\mathbf{y}$	

by (6.5). As the model corresponding to H_1 is more general, it is more likely to fit the data well so we assume that D_1 has the central distribution $\chi^2(N-p)$. On the other hand, D_0 may have a non-central distribution $\chi^2(N-q,v)$ if H_0 is not correct – see Section 5.6. In this case, $\triangle D = D_0 - D_1$ would have the non-central distribution $\chi^2(p-q,v)$ (provided appropriate conditions are satisfied – see Section 1.5). Therefore the statistic

$$F=\frac{D_0-D_1}{p-q}/\frac{D_1}{N-p}=\frac{\left(\mathbf{b}_0^T\mathbf{X}_0^T\mathbf{y}-\mathbf{b}_1^T\mathbf{X}_1^T\mathbf{y}\right)}{p-q}\bigg/\frac{(\mathbf{y}^T\mathbf{y}-\mathbf{b}_1^T\mathbf{X}_1^T\mathbf{y})}{N-p}$$

will have the central distribution $F(p-q,N-p)$ if H_0 is correct or F will otherwise have a non-central distribution. Therefore values of F that are large relative to the distribution $F(p-q,N-p)$ provide evidence against H_0 (see Figure 2.5).

This hypothesis test is often summarized by the Analysis of Variance table shown in Table 6.1.

6.2.5 Orthogonality

Usually inferences about a parameter for one explanatory variable depend on which other explanatory variables are included in the model. An exception is when the design matrix can be partitioned into components $\mathbf{X}_1,...,\mathbf{X}_m$ corresponding to submodels of interest,

$$\mathbf{X}=[\mathbf{X}_1,...,\mathbf{X}_m] \quad \text{for } m\leq p,$$

where $\mathbf{X}_j^T\mathbf{X}_k=\mathbf{O}$, a matrix of zeros, for each $j\neq k$. In this case, $\mathbf{X}$ is said to be **orthogonal**. Let $\boldsymbol{\beta}$ have corresponding components $\boldsymbol{\beta}_1,...,\boldsymbol{\beta}_m$ so that

$$\mathrm{E}(\mathbf{y})=\mathbf{X}\boldsymbol{\beta}=\mathbf{X}_1\boldsymbol{\beta}_1+\mathbf{X}_2\boldsymbol{\beta}_2+...+\mathbf{X}_m\boldsymbol{\beta}_m.$$

Typically, the components correspond to individual covariates or groups of associated explanatory variables such as dummy variables denoting levels of a factor. If $\mathbf{X}$ can be partitioned in this way then $\mathbf{X}^T\mathbf{X}$ is a block diagonal

Table 6.2 *Multiple hypothesis tests when the design matrix* $\mathbf{X}$ *is orthogonal.*

Source of variance	Degrees of freedom	Sum of squares
Model corresponding to H_1	p_1	$\mathbf{b}_1^T\mathbf{X}_1^T\mathbf{y}$
$\vdots$	$\vdots$	$\vdots$
Model corresponding to H_m	p_m	$\mathbf{b}_m^T\mathbf{X}_m^T\mathbf{y}$
Residual	$N - \sum_{j=1}^m p_j$	$\mathbf{y}^T\mathbf{y} - \mathbf{b}^T\mathbf{X}^T\mathbf{y}$
Total	N	$\mathbf{y}^T\mathbf{y}$

matrix

$$\mathbf{X}^T\mathbf{X} = \begin{bmatrix} \mathbf{X}_1^T\mathbf{X}_1 & & \mathbf{O} \\ & \ddots & \\ \mathbf{O} & & \mathbf{X}_m^T\mathbf{X}_m \end{bmatrix}. \quad \text{Also} \quad \mathbf{X}^T\mathbf{y} = \begin{bmatrix} \mathbf{X}_1^T\mathbf{y} \\ \vdots \\ \mathbf{X}_m^T\mathbf{y} \end{bmatrix}.$$

Therefore the estimates $\mathbf{b}_j = (\mathbf{X}_j^T\mathbf{X}_j)^{-1}\mathbf{X}_j^T\mathbf{y}$ are unaltered by the inclusion of other elements in the model and also

$$\mathbf{b}^T\mathbf{X}^T\mathbf{y} = \mathbf{b}_1^T\mathbf{X}_1^T\mathbf{y} + ... + \mathbf{b}_m^T\mathbf{X}_m^T\mathbf{y}.$$

Consequently, the hypotheses

$$H_1 : \boldsymbol{\beta}_1 = \mathbf{0}, ..., H_m : \boldsymbol{\beta}_m = \mathbf{0}$$

can be tested independently as shown in Table 6.2.

In practice, except for some well-designed experiments, the design matrix $\mathbf{X}$ is hardly ever orthogonal. Therefore inferences about any subset of parameters, $\boldsymbol{\beta}_j$ say, depend on the order in which other terms are included in the model. To overcome this ambiguity many statistical programs provide tests based on all other terms being included before $\mathbf{X}_j\boldsymbol{\beta}_j$ is added. The resulting sums of squares and hypothesis tests are sometimes called **Type III tests** (if the tests depend on the sequential order of fitting terms they are called Type I).

6.2.6 Residuals

Corresponding to the model formulation (6.2), the residuals are defined as

$$\widehat{e}_i = y_i - \mathbf{x}_i^T\mathbf{b} = y_i - \widehat{\mu}_i$$

where $\widehat{\mu}_i$ is the fitted value. The variance-covariance matrix of the vector of residuals $\widehat{\mathbf{e}}$ is

$$\begin{aligned} \mathrm{E}(\widehat{\mathbf{e}}\widehat{\mathbf{e}}^T) &= \mathrm{E}[(\mathbf{y} - \mathbf{X}\mathbf{b})(\mathbf{y} - \mathbf{X}\mathbf{b})^T] \\ &= \mathrm{E}\left(\mathbf{y}\mathbf{y}^T\right) - \mathbf{X}\mathrm{E}\left(\mathbf{b}\mathbf{b}^T\right)\mathbf{X}^T \\ &= \sigma^2\left[\mathbf{I} - \mathbf{X}(\mathbf{X}^T\mathbf{X})^{-1}\mathbf{X}^T\right] \end{aligned}$$

where $\mathbf{I}$ is the unit matrix. So the **standardized residuals** are

$$r_i = \frac{\widehat{e}_i}{\widehat{\sigma}(1 - h_{ii})^{1/2}}$$

where h_{ii} is the ith element on the diagonal of the **projection** or **hat matrix** $\mathbf{H} = \mathbf{X}(\mathbf{X}^T\mathbf{X})^{-1}\mathbf{X}^T$ and $\widehat{\sigma}^2$ is an estimate of σ^2.

These residuals should be used to check the adequacy of the fitted model using the various plots and other methods discussed in Section 2.3.4. These diagnostic tools include checking linearity of relationships between variables, serial independence of observations, Normality of residuals, and associations with other potential explanatory variables that are not included in the model.

6.2.7 Other diagnostics

In addition to residuals, there are numerous other methods to assess the adequacy of a model and to identify unusual or influential observations.

An **outlier** is an observation which is not well fitted by the model. An **influential observation** is one which has a relatively large effect on inferences based on the model. Influential observations may or may not be outliers and vice versa.

The value h_{ii}, the ith element on the diagonal of the hat matrix, is called the **leverage** of the ith observation. An observation with high leverage can make a substantial difference to the fit of the model. As a rule of thumb, if h_{ii} is greater than two or three times p/N it may be a concern (where p is the number of parameters and N the number of observations).

Measures which combine standardized residuals and leverage include

$$\mathrm{DFITS}_i = r_i \left(\frac{h_{ii}}{1 - h_{ii}}\right)^{1/2}$$

and **Cook's distance**

$$D_i = \frac{1}{p}\left(\frac{h_{ii}}{1 - h_{ii}}\right) r_i^2.$$

Large values of these statistics indicate that the ith observation is influential. Details of hypothesis tests for these and related statistics are given, for example, by Cook and Weisberg (1999).

Another approach to identifying influential observations is to fit a model with and without each observation and see what difference this makes to the estimates $\mathbf{b}$ and the overall goodness of fit statistics such as the deviance or the minimum value of the sum of squares criterion. For example, the statistic **delta-beta** is defined by

$$\triangle_i \widehat{\beta}_j = b_j - b_{j(i)}$$

where $b_{j(i)}$ denotes the estimate of β_j obtained when the ith observation is omitted from the data. These statistics can be standardized by dividing by their standard errors, and then they can be compared with the standard Normal distribution to identify unusually large ones. They can be plotted against the observation numbers i so that the 'offending' observations can be easily identified.

The delta-betas can be combined over all parameters using

$$D_i = \frac{1}{p}\left(\mathbf{b} - \mathbf{b}_{(i)}\right)^T \mathbf{X}^T\mathbf{X}(\mathbf{b} - \mathbf{b}_{(i)})$$

where $\mathbf{b}_{(i)}$ denotes the vector of estimates $b_{j(i)}$. This statistic is, in fact, equal to the Cook's distance (Neter et al., 1996).

Similarly the influence of the ith observation on the deviance, called **delta-deviance**, can be calculated as the difference between the deviance for the model fitted from all the data and the deviance for the same model with the ith observation omitted.

For Normal linear models there are algebraic simplifications of these statistics which mean that, in fact, the models do not have to be refitted omitting one observation at a time. The statistics can be calculated easily and are provided routinely be most statistical software. An overview of these diagnostic tools is given by the article by Chatterjee and Hadi (1986).

Once an influential observation or an outlier is detected, the first step is to determine whether it might be a measurement error, transcription error or some other mistake. It should it be removed from the data set only if there is a good substantive reason for doing so. Otherwise a possible solution is to retain it and report the results that are obtained with and without its inclusion in the calculations.

6.3 Multiple linear regression

If the explanatory variables are all continuous, the design matrix has a column of ones, corresponding to an intercept term in the linear component, and all the other elements are observed values of the explanatory variables. Multiple linear regression is the simplest Normal linear model for this situation. The following example provides an illustration.

6.3.1 Carbohydrate diet

The data in Table 6.3 show responses, percentages of total calories obtained from complex carbohydrates, for twenty male insulin-dependent diabetics who had been on a high-carbohydrate diet for six months. Compliance with the regime was thought to be related to age (in years), body weight (relative to

Table 6.3 *Carbohydrate, age, relative weight and protein for twenty male insulin-dependent diabetics; for units, see text (data from K. Webb, personal communication).*

Carbohydrate y	Age x_1	Weight x_2	Protein x_3
33	33	100	14
40	47	92	15
37	49	135	18
27	35	144	12
30	46	140	15
43	52	101	15
34	62	95	14
48	23	101	17
30	32	98	15
38	42	105	14
50	31	108	17
51	61	85	19
30	63	130	19
36	40	127	20
41	50	109	15
42	64	107	16
46	56	117	18
24	61	100	13
35	48	118	18
37	28	102	14

'ideal' weight for height) and other components of the diet, such as the percentage of calories as protein. These other variables are treated as explanatory variables.

We begin by fitting the model

$$\mathrm{E}(Y_i) = \mu_i = \beta_0 + \beta_1 x_{i1} + \beta_2 x_{i2} + \beta_3 x_{i3} \quad ; \quad Y_i \sim N(\mu_i, \sigma^2) \tag{6.6}$$

in which carbohydrate Y is linearly related to age x_1, relative weight x_2 and protein x_3 ($i = 1, ..., N = 20$). In this case

$$\mathbf{y} = \begin{bmatrix} Y_1 \\ \vdots \\ Y_N \end{bmatrix}, \quad \mathbf{X} = \begin{bmatrix} 1 & x_{11} & x_{12} & x_{13} \\ \vdots & \vdots & \vdots & \vdots \\ 1 & x_{N1} & x_{N2} & x_{N3} \end{bmatrix} \quad \text{and} \quad \boldsymbol{\beta} = \begin{bmatrix} \beta_0 \\ \vdots \\ \beta_3 \end{bmatrix}.$$

For these data

$$\mathbf{X}^T\mathbf{y} = \begin{bmatrix} 752 \\ 34596 \\ 82270 \\ 12105 \end{bmatrix}$$

and

$$\mathbf{X}^T\mathbf{X} = \begin{bmatrix} 20 & 923 & 2214 & 318 \\ 923 & 45697 & 102003 & 14780 \\ 2214 & 102003 & 250346 & 35306 \\ 318 & 14780 & 35306 & 5150 \end{bmatrix}.$$

Therefore the solution of $\mathbf{X}^T\mathbf{X}\mathbf{b} = \mathbf{X}^T\mathbf{y}$ is

$$\mathbf{b} = \begin{bmatrix} 36.9601 \\ -0.1137 \\ -0.2280 \\ 1.9577 \end{bmatrix}$$

and

$$(\mathbf{X}^T\mathbf{X})^{-1} = \begin{bmatrix} 4.8158 & -0.0113 & -0.0188 & -0.1362 \\ -0.0113 & 0.0003 & 0.0000 & -0.0004 \\ -0.0188 & 0.0000 & 0.0002 & -0.0002 \\ -0.1362 & -0.0004 & -0.0002 & 0.0114 \end{bmatrix}$$

correct to four decimal places. Also $\mathbf{y}^T\mathbf{y} = 29368$, $N\overline{y}^2 = 28275.2$ and $\mathbf{b}^T\mathbf{X}^T\mathbf{y} = 28800.337$. Using (6.4) to obtain an unbiased estimator of σ^2 we get $\widehat{\sigma}^2 = 35.479$ and hence we obtain the standard errors for elements of $\mathbf{b}$ which are shown in Table 6.4.

Table 6.4 *Estimates for model (6.6).*

Term	Estimate b_j	Standard error*
Constant	36.960	13.071
Coefficient for age	-0.114	0.109
Coefficient for weight	-0.228	0.083
Coefficient for protein	1.958	0.635

*Values calculated using more significant figures for $(\mathbf{X}^T\mathbf{X})^{-1}$ than shown above.

To illustrate the use of the deviance we test the hypothesis, H_0, that the response does not depend on age, i.e., $\beta_1 = 0$. The corresponding model is

$$\mathrm{E}(Y_i) = \beta_0 + \beta_2 x_{i2} + \beta_3 x_{i3}. \tag{6.7}$$

The matrix $\mathbf{X}$ for this model is obtained from the previous one by omitting the second column so that

$$\mathbf{X}^T\mathbf{y} = \begin{bmatrix} 752 \\ 82270 \\ 12105 \end{bmatrix}, \quad \mathbf{X}^T\mathbf{X} = \begin{bmatrix} 20 & 2214 & 318 \\ 2214 & 250346 & 35306 \\ 318 & 35306 & 5150 \end{bmatrix}$$

and hence

$$\mathbf{b} = \begin{bmatrix} 33.130 \\ -0.222 \\ 1.824 \end{bmatrix}.$$

For model (6.7), $\mathbf{b}^T\mathbf{X}^T\mathbf{y} = 28761.978$. The significance test for H_0 is summarized in Table 6.5. The value $F = 38.36/35.48 = 1.08$ is not significant compared with the $F(1, 16)$ distribution so the data provide no evidence against H_0, i.e., the response appears to be unrelated to age.

Table 6.5 *Analysis of Variance table comparing models (6.6) and (6.7).*

Source variation	Degrees of freedom	Sum of squares	Mean square
Model (6.7)	3	28761.978	
Improvement due to model (6.6)	1	38.359	38.36
Residual	16	567.663	35.48
Total	20	29368.000	

Notice that the parameter estimates for models (6.6) and (6.7) differ; for example, the coefficient for protein is 1.958 for the model including a term for age but 1.824 when the age term is omitted. This is an example of lack of orthogonality. It is illustrated further in Exercise 6.3(c) as the ANOVA table for testing the hypothesis that the coefficient for age is zero when both weight and protein are in the model, Table 6.5, differs from the ANOVA table when weight is not included.

6.3.2 Coefficient of determination, R^2

A commonly used measure of goodness of fit for multiple linear regression models is based on a comparison with the simplest or **minimal model** using the least squares criterion (in contrast to the maximal model and the log likelihood function which are used to define the deviance). For the model specified in (6.2), the least squares criterion is

$$S = \sum_{i=1}^{N} e_i^2 = \mathbf{e}^T\mathbf{e} = (\mathbf{Y} - \mathbf{X}\boldsymbol{\beta})^T (\mathbf{Y} - \mathbf{X}\boldsymbol{\beta})$$

and, from Section 6.2.2, the least squares estimate is $\mathbf{b} = (\mathbf{X}^T\mathbf{X})^{-1}\mathbf{X}^T\mathbf{y}$ so the minimum value of S is

$$\widehat{S} = (\mathbf{y} - \mathbf{X}\mathbf{b})^T (\mathbf{y} - \mathbf{X}\mathbf{b}) = \mathbf{y}^T\mathbf{y} - \mathbf{b}^T\mathbf{X}^T\mathbf{y}.$$

The simplest model is $\mathrm{E}(Y_i) = \mu$ for all i. In this case, $\boldsymbol{\beta}$ has the single element μ and $\mathbf{X}$ is a vector of N ones. So $\mathbf{X}^T\mathbf{X} = N$ and $\mathbf{X}^T\mathbf{y} = \sum y_i$ so that $\mathbf{b} = \widehat{\mu} = \overline{y}$. In this case, the value of S is

$$\widehat{S}_0 = \mathbf{y}^T\mathbf{y} - N\overline{y}^2 = \sum (y_i - \overline{y})^2 .$$

So $\widehat{S}_0$ is proportional to the variance of the observations and it is the largest or 'worst possible' value of S. The relative improvement in fit for any other model is

$$R^2 = \frac{\widehat{S}_0 - \widehat{S}}{\widehat{S}_0} = \frac{\mathbf{b}^T\mathbf{X}^T\mathbf{y} - N\overline{y}^2}{\mathbf{y}^T\mathbf{y} - N\overline{y}^2}.$$

R^2 is called the **coefficient of determination**. It can be interpreted as the proportion of the total variation in the data which is explained by the model.

For example, for the carbohydrate data $R^2 = 0.48$ for model (6.5), so 48% of the variation is 'explained' by the model. If the term for age is dropped, for model (6.6) $R^2 = 0.445$, so 44.5% of variation is 'explained'.

If the model does not fit the data much better than the minimal model then $\widehat{S}$ will be almost equal to $\widehat{S}_0$ and R^2 will be almost zero. On the other hand if the maximal model is fitted, with one parameter μ_i for each observation Y_i, then $\boldsymbol{\beta}$ has N elements, $\mathbf{X}$ is the $N \times N$ unit matrix $\mathbf{I}$ and $\mathbf{b} = \mathbf{y}$ (i.e., $\widehat{\mu}_i = y_i$). So for the maximal model $\mathbf{b}^T\mathbf{X}^T\mathbf{y} = \mathbf{y}^T\mathbf{y}$ and hence $\widehat{S} = 0$ and $R^2 = 1$, corresponding to a 'perfect' fit. In general, $0 < R^2 < 1$. The square root of R^2 is called the **multiple correlation coefficient**.

Despite its popularity and ease of interpretation R^2 has limitations as a measure of goodness of fit. Its sampling distribution is not readily determined. Also it always increases as more parameters are added to the model, so modifications of R^2 have to be used to adjust for the number of parameters.

6.3.3 Model selection

Many applications of multiple linear regression involve numerous explanatory variables and it is important to identify a subset of these variables that provides a good, yet parsimonious, model for the response. The usual procedure is to add or delete terms sequentially from the model; this is called **stepwise regression**. Details of the methods are given in standard textbooks on regression such as Draper and Smith (1998) or Neter et al. (1996).

If some of the explanatory variables are highly correlated with one another, this is called **collinearity** or **multicollinearity**. This condition has several undesirable consequences. Firstly, the columns of the design matrix $\mathbf{X}$ may be nearly linearly dependent so that $\mathbf{X}^T\mathbf{X}$ is nearly singular and the estimating equation $\left(\mathbf{X}^T\mathbf{X}\right)\mathbf{b} = \mathbf{X}^T\mathbf{y}$ is ill-conditioned. This means that the solution $\mathbf{b}$ will be unstable in the sense that small changes in the data may cause large charges in $\mathbf{b}$ (see Section 6.2.7). Also at least some of the elements of $\sigma^2(\mathbf{X}^T\mathbf{X})^{-1}$ will be large giving large variances or covariances for elements of $\mathbf{b}$. Secondly, collinearity means that choosing the best subset of explanatory variables may be difficult.

Collinearity can be detected by calculating the **variance inflation factor** for each explanatory variable

$$\text{VIF}_j = \frac{1}{1 - R^2_{(j)}}$$

where $R^2_{(j)}$ is the coefficient of determination obtained from regressing the jth explanatory variable against all the other explanatory variables. If it is uncorrelated with all the others then VIF = 1. VIF increases as the correlation increases. It is suggest, by Montgomery and Peck (1992) for example, that one should be concerned if VIF > 5.

If several explanatory variables are highly correlated it may be impossible, on statistical grounds alone, to determine which one should be included in the model. In this case extra information from the substantive area from which the data came, an alternative specification of the model or some other non-computational approach may be needed.

6.4 Analysis of variance

Analysis of variance is the term used for statistical methods for comparing means of groups of continuous observations where the groups are defined by the levels of factors. In this case all the explanatory variables are categorical and all the elements of the design matrix **X** are dummy variables. As illustrated in Example 2.4.3, the choice of dummy variables is, to some extent, arbitrary. An important consideration is the optimal choice of specification of **X**. The major issues are illustrated by two numerical examples with data from two (fictitious) designed experiments.

6.4.1 One factor analysis of variance

The data in Table 6.6 are similar to the plant weight data in Exercise 2.1. An experiment was conducted to compare yields Y_i (measured by dried weight of plants) under a control condition and two different treatment conditions. Thus the response, dried weight, depends on one factor, growing condition, with three levels. We are interested in whether the response means differ among the groups.

More generally, if experimental units are randomly allocated to groups corresponding to J levels of a factor, this is called a **completely randomized experiment**. The data can be set out as shown in Table 6.7.

The responses at level j, $Y_{j1}, ..., Y_{jn_j}$, all have the same expected value and so they are called **replicates**. In general there may be different numbers of observations n_j at each level.

To simplify the discussion suppose all the groups have the same sample size so $n_j = K$ for $j = 1, ..., J$. The response **y** is the column vector of all $N = JK$ measurements

$$\mathbf{y} = [Y_{11}, Y_{12}, ..., Y_{1K}, Y_{21}, ..., Y_{2K}, ..., Y_{J1}, ..., Y_{JK}]^T.$$

We consider three different specifications of a model to test the hypothesis that the response means differ among the factor levels.

(a) The simplest specification is

$$\mathrm{E}(Y_{jk}) = \mu_j \quad \text{for } j = 1, ..., K. \tag{6.8}$$

Table 6.6 *Dried weights y_i of plants from three different growing conditions.*

	Control	Treatment A	Treatment B
	4.17	4.81	6.31
	5.58	4.17	5.12
	5.18	4.41	5.54
	6.11	3.59	5.50
	4.50	5.87	5.37
	4.61	3.83	5.29
	5.17	6.03	4.92
	4.53	4.89	6.15
	5.33	4.32	5.80
	5.14	4.69	5.26
$\sum y_i$	50.32	46.61	55.26
$\sum y_i^2$	256.27	222.92	307.13

Table 6.7 *Data from a completely randomized experiment with J levels of a factor A.*

	Factor level			
	A_1	A_2	$\cdots$	A_J
	Y_{11}	Y_{21}		Y_{J1}
	Y_{12}	Y_{22}		Y_{J2}
	$\vdots$			$\vdots$
	Y_{1n_1}	Y_{2n_2}		Y_{Jn_J}
Total	$Y_{1.}$	$Y_{2.}$	$\cdots$	$Y_{J.}$

This can be written as

$$\mathrm{E}(Y_i) = \sum_{j=1}^{J} x_{ij}\mu_j, \quad i = 1, ..., N$$

where $x_{ij} = 1$ if response Y_i corresponds to level A_j and $x_{ij} = 0$ otherwise. Thus, $\mathrm{E}(\mathbf{y}) = \mathbf{X}\boldsymbol{\beta}$ with

$$\boldsymbol{\beta} = \begin{bmatrix} \mu_1 \\ \mu_2 \\ \vdots \\ \mu_J \end{bmatrix} \quad \text{and} \quad \mathbf{X} = \begin{bmatrix} \mathbf{1} & \mathbf{0} & \cdots & \mathbf{0} \\ \mathbf{0} & \mathbf{1} & & \vdots \\ \vdots & . & \mathbf{O} & \\ & \mathbf{O} & . & \mathbf{0} \\ \mathbf{0} & & & \mathbf{1} \end{bmatrix}$$

where $\mathbf{0}$ and $\mathbf{1}$ are vectors of length K of zeros and ones respectively, and $\mathbf{O}$

indicates that the remaining terms of the matrix are all zeros. Then $\mathbf{X}^T\mathbf{X}$ is the $J \times J$ diagonal matrix

$$\mathbf{X}^T\mathbf{X} = \begin{bmatrix} K & & & & \\ & \ddots & & \mathbf{O} & \\ & & K & & \\ & & & \ddots & \\ & \mathbf{O} & & & K \end{bmatrix} \quad \text{and} \quad \mathbf{X}^T\mathbf{y} = \begin{bmatrix} Y_{1.} \\ Y_{2.} \\ \vdots \\ Y_{J.} \end{bmatrix}.$$

So from equation (6.3)

$$\mathbf{b} = \frac{1}{K}\begin{bmatrix} Y_{1.} \\ Y_{2.} \\ \vdots \\ Y_{J.} \end{bmatrix} = \begin{bmatrix} \overline{Y}_1 \\ \overline{Y}_2 \\ \vdots \\ \overline{Y}_J \end{bmatrix}$$

and

$$\mathbf{b}^T\mathbf{X}^T\mathbf{y} = \frac{1}{K}\sum_{j=1}^{J} Y_{j.}^2 \, .$$

The fitted values are $\widehat{\mathbf{y}} = [\overline{y}_1, \overline{y}_1, ..., \overline{y}_1, \overline{y}_2, ..., \overline{y}_J]^T$. The disadvantage of this simple formulation of the model is that it cannot be extended to more than one factor. To generalize further, we need to specify the model so that parameters for levels and combinations of levels of factors reflect differential effects beyond some average or specified response.

(b) The second model is one such formulation:

$$E(Y_{jk}) = \mu + \alpha_j, \; j = 1, ..., J$$

where μ is the average effect for all levels and α_j is an additional effect due to level A_j. For this parameterization there are $J+1$ parameters.

$$\boldsymbol{\beta} = \begin{bmatrix} \mu \\ \alpha_1 \\ \vdots \\ \alpha_J \end{bmatrix}, \quad \mathbf{X} = \begin{bmatrix} \mathbf{1} & \mathbf{1} & \mathbf{0} & \cdots & \mathbf{0} \\ \mathbf{1} & \mathbf{0} & \mathbf{1} & & \\ \vdots & & & \mathbf{O} & \\ \vdots & \mathbf{O} & & & \\ \mathbf{1} & & & & \mathbf{1} \end{bmatrix}$$

where $\mathbf{0}$ and $\mathbf{1}$ are vectors of length K and $\mathbf{O}$ denotes a matrix of zeros. Thus

$$\mathbf{X}^T\mathbf{y} = \begin{bmatrix} Y_{..} \\ Y_{1.} \\ \vdots \\ Y_{J.} \end{bmatrix} \quad \text{and} \quad \mathbf{X}^T\mathbf{X} = \begin{bmatrix} N & K & \dots & K \\ K & K & & \\ \vdots & & \mathbf{O} & \\ \vdots & \mathbf{O} & & \\ K & & & K \end{bmatrix}.$$

The first row (or column) of the $(J+1)\times(J+1)$ matrix $\mathbf{X}^T\mathbf{X}$ is the sum of the remaining rows (or columns) so $\mathbf{X}^T\mathbf{X}$ is singular and there is no unique solution of the normal equations $\mathbf{X}^T\mathbf{X}\mathbf{b} = \mathbf{X}^T\mathbf{y}$. The general solution can be written as

$$\mathbf{b} = \begin{bmatrix} \widehat{\mu} \\ \widehat{\alpha}_1 \\ \vdots \\ \widehat{\alpha}_J \end{bmatrix} = \frac{1}{K}\begin{bmatrix} 0 \\ Y_{1.} \\ \vdots \\ Y_{J.} \end{bmatrix} - \lambda \begin{bmatrix} -1 \\ 1 \\ \vdots \\ 1 \end{bmatrix}$$

where λ is an arbitrary constant. It is traditional to impose the additional **sum-to-zero constraint**

$$\sum_{j=1}^{J} \alpha_j = 0$$

so that

$$\frac{1}{K}\sum_{j=1}^{J} Y_{j.} - J\lambda = 0$$

and hence

$$\lambda = \frac{1}{JK}\sum_{j=1}^{J} Y_{j.} = \frac{Y_{..}}{N}\,.$$

This gives the solution

$$\widehat{\mu} = \frac{Y_{..}}{N} \quad \text{and} \quad \widehat{\alpha}_j = \frac{Y_{j.}}{K} - \frac{Y_{..}}{N} \quad \text{for } j = 1, ..., J.$$

Hence

$$\mathbf{b}^T\mathbf{X}^T\mathbf{y} = \frac{Y_{..}^2}{N} + \sum_{j=1}^{J} Y_{j.}\left(\frac{Y_{j.}}{K} - \frac{Y_{..}}{N}\right) = \frac{1}{K}\sum_{j=1}^{J} Y_{j.}^2$$

which is the same as for the first version of the model and the fitted values $\widehat{\mathbf{y}} = [\overline{y}_1, \overline{y}_1, ..., \overline{y}_J]^T$ are also the same. Sum-to-zero constraints are used in most standard statistical software.

(c) A third version of the model is $E(Y_{jk}) = \mu + \alpha_j$ with the constraint that $\alpha_1 = 0$. Thus μ represents the effect of the first level and α_j measures the difference between the first level and jth level of the factor. This is called a **corner-point parameterization**. For this version there are J parameters

$$\boldsymbol{\beta} = \begin{bmatrix} \mu \\ \alpha_2 \\ \vdots \\ \alpha_J \end{bmatrix}. \qquad \text{Also} \quad \mathbf{X} = \begin{bmatrix} \mathbf{1} & \mathbf{0} & \cdots & \mathbf{0} \\ \mathbf{1} & \mathbf{1} & & \\ \vdots & & \ddots & \mathbf{O} \\ \vdots & \mathbf{O} & & \\ \mathbf{1} & & & \mathbf{1} \end{bmatrix}$$

$$\text{so } \mathbf{X}^T\mathbf{y} = \begin{bmatrix} Y_{..} \\ Y_{2.} \\ \vdots \\ Y_{J.} \end{bmatrix} \quad \text{and} \quad \mathbf{X}^T\mathbf{X} = \begin{bmatrix} N & K & \dots & & K \\ K & K & & & \\ \vdots & & \ddots & \mathbf{O} & \\ \vdots & & \mathbf{O} & & \\ K & & & & K \end{bmatrix}.$$

The $J \times J$ matrix $\mathbf{X}^T\mathbf{X}$ is non-singular so there is a unique solution

$$\mathbf{b} = \frac{1}{K} \begin{bmatrix} Y_{1.} \\ Y_{2.} - Y_{1.} \\ \vdots \\ Y_{J.} - Y_{1.} \end{bmatrix}$$

Also $\mathbf{b}^T\mathbf{X}^T\mathbf{y} = \frac{1}{K}\left[Y_{..}Y_{1.} + \sum_{j=2}^{J} Y_{j.}(Y_{j.} - Y_{1.})\right] = \frac{1}{K}\sum_{j=1}^{J} Y_{j.}^2$ and the fitted values $\widehat{\mathbf{y}} = [\overline{y}_1, \overline{y}_1, ..., \overline{y}_J]^T$ are the same as before.

Thus, although the three specifications of the model differ, the value of $\mathbf{b}^T\mathbf{X}^T\mathbf{y}$ and hence

$$D_1 = \frac{1}{\sigma^2}\left(\mathbf{y}^T\mathbf{y} - \mathbf{b}^T\mathbf{X}^T\mathbf{y}\right) = \frac{1}{\sigma^2}\left[\sum_{j=1}^{J}\sum_{k=1}^{K} Y_{jk}^2 - \frac{1}{K}\sum_{j=1}^{J} Y_{j.}^2\right]$$

is the same in each case.

These three versions of the model all correspond to the hypothesis H_1 that the response means for each level may differ. To compare this with the null hypothesis H_0 that the means are all equal, we consider the model $\mathrm{E}(Y_{jk}) = \mu$ so that $\boldsymbol{\beta} = [\mu]$ and $\mathbf{X}$ is a vector of N ones. Then $\mathbf{X}^T\mathbf{X} = N, \mathbf{X}^T\mathbf{y} = Y_{..}$ and hence $\mathbf{b} = \widehat{\mu} = Y_{..}/N$ so that $\mathbf{b}^T\mathbf{X}^T\mathbf{y} = Y_{..}^2/N$ and

$$D_0 = \frac{1}{\sigma^2}\left[\sum_{j=1}^{J}\sum_{k=1}^{K} Y_{jk}^2 - \frac{Y_{..}^2}{N}\right].$$

To test H_0 against H_1 we assume that H_1 is correct so that $D_1 \sim \chi^2(N-J)$. If, in addition, H_0 is correct then $D_0 \sim \chi^2(N-1)$, otherwise D_0 has a non-central chi-squared distribution. Thus if H_0 is correct

$$D_0 - D_1 = \frac{1}{\sigma^2}\left[\frac{1}{K}\sum_{j=1}^{J} Y_{j.}^2 - \frac{1}{N}Y_{..}^2\right] \sim \chi^2(J-1)$$

and so

$$F = \frac{D_0 - D_1}{J-1} \bigg/ \frac{D_1}{N-J} \sim F(J-1, N-J).$$

If H_0 is not correct then F is likely to be larger than predicted from the distribution $F(J-1, N-J)$. Conventionally this hypothesis test is set out in an ANOVA table.

For the plant weight data

$$\frac{Y_{..}^2}{N} = 772.0599, \quad \frac{1}{K}\sum_{j=1}^{J} Y_{j.}^2 = 775.8262$$

so

$$D_0 - D_1 = 3.7663/\sigma^2$$

and

$$\sum_{j=1}^{J}\sum_{k=1}^{K} Y_{jk}^2 = 786.3183$$

so $D_1 = 10.4921/\sigma^2$. Hence the hypothesis test is summarized in Table 6.8.

Table 6.8 *ANOVA table for plant weight data in Table 6.6.*

Source of variation	Degrees of freedom	Sum of squares	Mean square	F
Mean	1	772.0599		
Between treatment	2	3.7663	1.883	4.85
Residual	27	10.4921	0.389	
Total	30	786.3183		

Since $F = 4.85$ is significant at the 5% level when compared with the $F(2, 27)$ distribution, we conclude that the group means differ.

To investigate this result further it is convenient to use the first version of the model (6.8), $\mathrm{E}(Y_{jk}) = \mu_j$. The estimated means are

$$\mathbf{b} = \begin{bmatrix} \widehat{\mu}_1 \\ \widehat{\mu}_2 \\ \widehat{\mu}_3 \end{bmatrix} = \begin{bmatrix} 5.032 \\ 4.661 \\ 5.526 \end{bmatrix}.$$

If we use the estimator

$$\widehat{\sigma}^2 = \frac{1}{N-J}(\mathbf{y} - \mathbf{Xb})^T(\mathbf{y} - \mathbf{Xb}) = \frac{1}{N-J}(\mathbf{y}^T\mathbf{y} - \mathbf{b}^T\mathbf{X}^T\mathbf{y})$$

(Equation 6.4), we obtain $\widehat{\sigma}^2 = 10.4921/27 = 0.389$ (i.e., the residual mean square in Table 6.8). The variance-covariance matrix of $\mathbf{b}$ is $\widehat{\sigma}^2\left(\mathbf{X}^T\mathbf{X}\right)^{-1}$ where

$$\mathbf{X}^T\mathbf{X} = \begin{bmatrix} 10 & 0 & 0 \\ 0 & 10 & 0 \\ 0 & 0 & 10 \end{bmatrix},$$

so the standard error of each element of $\mathbf{b}$ is $\sqrt{0.389/10} = 0.197$. Now it can be seen that the significant effect is due to the mean for treatment B,

$\widehat{\mu}_3 = 5.526$, being significantly (more than two standard deviations) larger than the other two means. Note that if several pairwise comparisons are made among elements of **b**, the standard errors should be adjusted to take account of multiple comparisons – see, for example, Neter et al. (1996).

6.4.2 Two factor analysis of variance

Consider the fictitious data in Table 6.9 in which factor A (with $J = 3$ levels) and factor B (with $K = 2$ levels) are **crossed** so that there are JK subgroups formed by all combinations of A and B levels. In each subgroup there are $L = 2$ observations or **replicates**.

Table 6.9 *Fictitious data for two-factor ANOVA with equal numbers of observations in each subgroup.*

Levels of factor A	Levels of factor B		
	B$_1$	B$_2$	Total
A$_1$	6.8, 6.6	5.3, 6.1	24.8
A$_2$	7.5, 7.4	7.2, 6.5	28.6
A$_3$	7.8, 9.1	8.8, 9.1	34.8
Total	45.2	43.0	88.2

The main hypotheses are:

H_I: there are no interaction effects, i.e., the effects of A and B are additive;

H_A: there are no differences in response associated with different levels of factor A;

H_B: there are no differences in response associated with different levels of factor B.

Thus we need to consider a **saturated model** and three **reduced models** formed by omitting various terms from the saturated model.

1. The saturated model is

$$\mathrm{E}(Y_{jkl}) = \mu + \alpha_j + \beta_k + (\alpha\beta)_{jk} \tag{6.9}$$

where the terms $(\alpha\beta)_{jk}$ correspond to **interaction effects** and α_j and β_k to **main effects** of the factors;

2. The **additive model** is

$$\mathrm{E}(Y_{jkl}) = \mu + \alpha_j + \beta_k. \tag{6.10}$$

This compared to the saturated model to test hypothesis H_I.

3. The model formed by omitting effects due to B is

$$\mathrm{E}(Y_{jkl}) = \mu + \alpha_j. \tag{6.11}$$

This is compared to the additive model to test hypothesis H_B.

4. The model formed by omitting effects due to A is

$$\mathrm{E}(Y_{jkl}) = \mu + \beta_k. \tag{6.12}$$

This is compared to the additive model to test hypothesis H_A.

The models (6.9) to (6.12) have too many parameters because replicates in the same subgroup have the same expected value so there can be at most JK independent expected values but the saturated model has $1 + J + K + JK = (J+1)(K+1)$ parameters. To overcome this difficulty (which leads to the singularity of $\mathbf{X}^T\mathbf{X}$), we can impose the extra constraints

$$\alpha_1 + \alpha_2 + \alpha_3 = 0, \quad \beta_1 + \beta_2 = 0,$$

$$(\alpha\beta)_{11} + (\alpha\beta)_{12} = 0, \quad (\alpha\beta)_{21} + (\alpha\beta)_{22} = 0, \quad (\alpha\beta)_{31} + (\alpha\beta)_{32} = 0,$$

$$(\alpha\beta)_{11} + (\alpha\beta)_{21} + (\alpha\beta)_{31} = 0$$

(the remaining condition $(\alpha\beta)_{12} + (\alpha\beta)_{22} + (\alpha\beta)_{32} = 0$ follows from the last four equations). These are the conventional sum-to-zero constraint equations for ANOVA. Alternatively, we can take

$$\alpha_1 = \beta_1 = (\alpha\beta)_{11} = (\alpha\beta)_{12} = (\alpha\beta)_{21} = (\alpha\beta)_{31} = 0$$

as the corner-point constraints. In either case the numbers of (linearly) independent parameters are: 1 for μ, $J-1$ for the α_j's, $K-1$ for the β_k's, and $(J-1)(K-1)$ for the $(\alpha\beta)_{jk}$'s, giving a total of JK parameters.

We will fit all four models using, for simplicity, the corner point constraints. The response vector is

$$\mathbf{y} = [6.8, 6.6, 5.3, 6.1, 7.5, 7.4, 7.2, 6.5, 7.8, 9.1, 8.8, 9.1]^T$$

and $\mathbf{y}^T\mathbf{y} = 664.1$.

For the saturated model (6.9) with constraints

$$\alpha_1 = \beta_1 = (\alpha\beta)_{11} = (\alpha\beta)_{12} = (\alpha\beta)_{21} = (\alpha\beta)_{31} = 0$$

$$\boldsymbol{\beta} = \begin{bmatrix} \mu \\ \alpha_2 \\ \alpha_3 \\ \beta_2 \\ (\alpha\beta)_{22} \\ (\alpha\beta)_{32} \end{bmatrix}, \quad \mathbf{X} = \begin{bmatrix} 100000 \\ 100000 \\ 100100 \\ 100100 \\ 110000 \\ 110000 \\ 110110 \\ 110110 \\ 101000 \\ 101000 \\ 101101 \\ 101101 \end{bmatrix}, \quad \mathbf{X}^T\mathbf{y} = \begin{bmatrix} Y_{...} \\ Y_{2..} \\ Y_{3..} \\ Y_{12.} \\ Y_{22.} \\ Y_{32.} \end{bmatrix} = \begin{bmatrix} 88.2 \\ 28.6 \\ 34.8 \\ 43.0 \\ 13.7 \\ 17.9 \end{bmatrix},$$

$$\mathbf{X}^T\mathbf{X} = \begin{bmatrix} 12 & 4 & 4 & 6 & 2 & 2 \\ 4 & 4 & 0 & 2 & 2 & 0 \\ 4 & 0 & 4 & 2 & 0 & 2 \\ 6 & 2 & 2 & 6 & 2 & 2 \\ 2 & 2 & 0 & 2 & 2 & 0 \\ 2 & 0 & 2 & 2 & 0 & 2 \end{bmatrix}, \quad \mathbf{b} = \begin{bmatrix} 6.7 \\ 0.75 \\ 1.75 \\ -1.0 \\ 0.4 \\ 1.5 \end{bmatrix}$$

and $\mathbf{b}^T\mathbf{X}^T\mathbf{y} = 662.62$.

For the additive model (6.10) with the constraints $\alpha_1 = \beta_1 = 0$ the design matrix is obtained by omitting the last two columns of the design matrix for the saturated model. Thus

$$\boldsymbol{\beta} = \begin{bmatrix} \mu \\ \alpha_2 \\ \alpha_3 \\ \beta_2 \end{bmatrix}, \quad \mathbf{X}^T\mathbf{X} = \begin{bmatrix} 12 & 4 & 4 & 6 \\ 4 & 4 & 0 & 2 \\ 4 & 0 & 4 & 2 \\ 6 & 2 & 2 & 6 \end{bmatrix}, \quad \mathbf{X}^T\mathbf{y} = \begin{bmatrix} 88.2 \\ 28.6 \\ 34.8 \\ 43.0 \end{bmatrix}$$

and hence

$$\mathbf{b} = \begin{bmatrix} 6.383 \\ 0.950 \\ 2.500 \\ -0.367 \end{bmatrix}$$

so that $\mathbf{b}^T\mathbf{X}^T\mathbf{y} = 661.4133$.

For model (6.11) omitting the effects of levels of factor B and using the constraint $\alpha_1 = 0$, the design matrix is obtained by omitting the last three columns of the design matrix for the saturated model. Therefore

$$\boldsymbol{\beta} = \begin{bmatrix} \mu \\ \alpha_2 \\ \alpha_3 \end{bmatrix}, \quad \mathbf{X}^T\mathbf{X} = \begin{bmatrix} 12 & 4 & 4 \\ 4 & 4 & 0 \\ 4 & 0 & 4 \end{bmatrix}, \quad \mathbf{X}^T\mathbf{y} = \begin{bmatrix} 88.2 \\ 28.6 \\ 34.8 \end{bmatrix}$$

and hence

$$\mathbf{b} = \begin{bmatrix} 6.20 \\ 0.95 \\ 2.50 \end{bmatrix}$$

so that $\mathbf{b}^T\mathbf{X}^T\mathbf{y} = 661.01$.

The design matrix for model (6.12) with constraint $\beta_1 = 0$ comprises the first and fourth columns of the design matrix for the saturated model. Therefore

$$\boldsymbol{\beta} = \begin{bmatrix} \mu \\ \beta_2 \end{bmatrix}, \quad \mathbf{X}^T\mathbf{X} = \begin{bmatrix} 12 & 6 \\ 6 & 6 \end{bmatrix}, \quad \mathbf{X}^T\mathbf{y} = \begin{bmatrix} 88.2 \\ 43.0 \end{bmatrix}$$

and hence

$$\mathbf{b} = \begin{bmatrix} 7.533 \\ -0.367 \end{bmatrix}$$

so that $\mathbf{b}^T\mathbf{X}^T\mathbf{y} = 648.6733$.

Finally for the model with only a mean effect $\mathrm{E}(Y_{jkl}) = \mu$, the estimate is $\mathbf{b} = [\widehat{\mu}] = 7.35$ and so $\mathbf{b}^T\mathbf{X}^T\mathbf{y} = 648.27$.

The results of these calculations are summarized in Table 6.10. The subscripts S, I, A, B and M refer to the saturated model, models corresponding to H_I, H_A and H_B and the model with only the overall mean, respectively. The scaled deviances are the terms $\sigma^2 D = \mathbf{y}^T\mathbf{y} - \mathbf{b}^T\mathbf{X}^T\mathbf{y}$. The degrees of freedom, d.f., are given by N minus the number of parameters in the model.

Table 6.10 *Summary of calculations for data in Table 6.9.*

Model	d.f.	$\mathbf{b}^T\mathbf{X}^T\mathbf{y}$	Scaled Deviance
$\mu + \alpha_j + \beta_k + (\alpha\beta)_{jk}$	6	662.6200	$\sigma^2 D_S = 1.4800$
$\mu + \alpha_j + \beta_k$	8	661.4133	$\sigma^2 D_I = 2.6867$
$\mu + \alpha_j$	9	661.0100	$\sigma^2 D_B = 3.0900$
$\mu + \beta_k$	10	648.6733	$\sigma^2 D_A = 15.4267$
μ	11	648.2700	$\sigma^2 D_M = 15.8300$

To test H_I we assume that the saturated model is correct so that $D_S \sim \chi^2(6)$. If H_I is also correct then $D_I \sim \chi^2(8)$ so that $D_I - D_S \sim \chi^2(2)$ and

$$F = \frac{D_I - D_S}{2} \bigg/ \frac{D_S}{6} \sim F(2,6).$$

The value of

$$F = \frac{2.6867 - 1.48}{2\sigma^2} \bigg/ \frac{1.48}{6\sigma^2} = 2.45$$

is not statistically significant so the data do not provide evidence against H_I. Since H_I is not rejected we proceed to test H_A and H_B. For H_B we consider the difference in fit between the models (6.10) and (6.11) i.e., $D_B - D_I$ and compare this with D_S using

$$F = \frac{D_B - D_I}{1} \bigg/ \frac{D_S}{6} = \frac{3.09 - 2.6867}{\sigma^2} \bigg/ \frac{1.48}{6\sigma^2} = 1.63$$

which is not significant compared to the $F(1,6)$ distribution, suggesting that there are no differences due to levels of factor B. The corresponding test for H_A gives $F = 25.82$ which is significant compared with $F(2,6)$ distribution. Thus we conclude that the response means are affected only by differences in the levels of factor A. The most appropriate choice for the denominator for the F ratio, D_S or D_I, is debatable. D_S comes from a more complex model and is more likely to correspond to a central chi-squared distribution, but it has fewer degrees of freedom.

The ANOVA table for these data is shown in Table 6.11. The first number in the sum of squares column is the value of $\mathbf{b}^T\mathbf{X}^T\mathbf{y}$ corresponding to the simplest model $\mathrm{E}(Y_{jkl}) = \mu$.

A feature of these data is that the hypothesis tests are independent in the

Table 6.11 *ANOVA table for data in Table 6.8.*

Source of variation	Degrees of freedom	Sum of squares	Mean square	F
Mean	1	648.2700		
Levels of A	2	12.7400	6.3700	25.82
Levels of B	1	0.4033	0.4033	1.63
Interactions	2	1.2067	0.6033	2.45
Residual	6	1.4800	0.2467	
Total	12	664.1000		

sense that the results are not affected by which terms – other than those relating to the hypothesis in question – are also in the model. For example, the hypothesis of no differences due to factor B, $H_B : \beta_k = 0$ for all k, could equally well be tested using either models $\mathrm{E}(Y_{jkl}) = \mu + \alpha_j + \beta_k$ and $\mathrm{E}(Y_{jkl}) = \mu + \alpha_j$ and hence

$$\sigma^2 D_B - \sigma^2 D_I = 3.0900 - 2.6867 = 0.4033,$$

or models

$$E(Y_{jkl}) = \mu + \beta_k \quad \text{and} \quad E(Y_{jkl}) = \mu$$

and hence

$$\sigma^2 D_M - \sigma^2 D_A = 15.8300 - 15.4267 = 0.4033.$$

The reason is that the data are **balanced**, that is, there are equal numbers of observations in each subgroup. For balanced data it is possible to specify the design matrix in such a way that it is orthogonal (see Section 6.2.5 and Exercise 6.7). An example in which the hypothesis tests are not independent is given in Exercise 6.8.

The estimated sample means for each subgroup can be calculated from the values of $\mathbf{b}$. For example, for the saturated model (6.9) the estimated mean of the subgroup with the treatment combination A_3 and B_2 is $\widehat{\mu} + \widehat{\alpha}_3 + \widehat{\beta}_2 + \widehat{(\alpha\beta)}_{32} = 6.7 + 1.75 - 1.0 + 1.5 = 8.95$.

The estimate for the same mean from the additive model (6.10) is

$$\widehat{\mu} + \widehat{\alpha}_3 + \widehat{\beta}_2 = 6.383 + 2.5 - 0.367 = 8.516.$$

This shows the importance of deciding which model to use to summarize the data.

To assess the adequacy of an ANOVA model, residuals should be calculated and examined for unusual patterns, Normality, independence, and so on, as described in Section 6.2.6.

6.5 Analysis of covariance

Analysis of covariance is the term used for models in which some of the explanatory variables are dummy variables representing factor levels and others are continuous measurements, called **covariates**. As with ANOVA, we are interested in comparing means of subgroups defined by factor levels but, recognizing that the covariates may also affect the responses, we compare the means after 'adjustment' for covariate effects.

A typical example is provided by the data in Table 6.12. The responses Y_{jk} are achievement scores measured at three levels of a factor representing three different training methods, and the covariates x_{jk} are aptitude scores measured before training commenced. We want to compare the training methods, taking into account differences in initial aptitude between the three groups of subjects.

The data are plotted in Figure 6.1. There is evidence that the achievement scores y increase linearly with aptitude x and that the y values are generally higher for training groups B and C than for A.

Table 6.12 *Achievement scores (data from Winer, 1971, p. 776.)*

	Training method					
	A		B		C	
	y	x	y	x	y	x
	6	3	8	4	6	3
	4	1	9	5	7	2
	5	3	7	5	7	2
	3	1	9	4	7	3
	4	2	8	3	8	4
	3	1	5	1	5	1
	6	4	7	2	7	4
Total	31	15	53	24	47	19
Sums of squares	147	41	413	96	321	59
$\sum xy$	75		191		132	

To test the hypothesis that there are no differences in mean achievement scores among the three training methods, after adjustment for initial aptitude, we compare the saturated model

$$\mathrm{E}(Y_{jk}) = \mu_j + \gamma x_{jk} \tag{6.13}$$

with the reduced model

$$\mathrm{E}(Y_{jk}) = \mu + \gamma x_{jk} \tag{6.14}$$

where $j = 1$ for method A, $j = 2$ for method B and $j = 3$ for method C, and

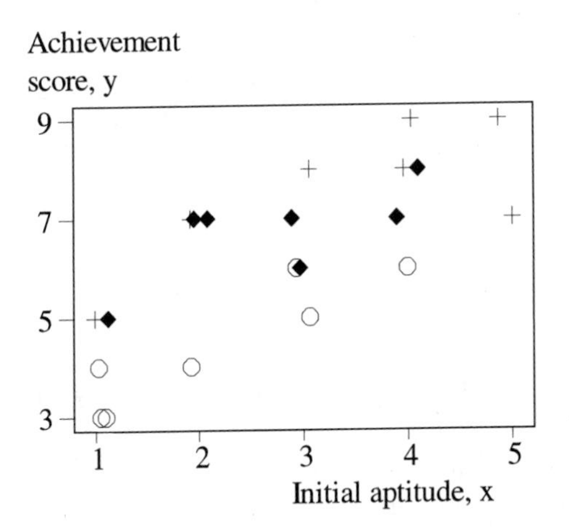

Figure 6.1 *Achievement and initial aptitude scores: circles denote training method A, crosses denote method B and diamonds denote method C.*

$k = 1, ..., 7$. Let

$$\mathbf{y}_j = \begin{bmatrix} Y_{j1} \\ \vdots \\ Y_{j7} \end{bmatrix} \quad \text{and} \quad \mathbf{x}_j = \begin{bmatrix} x_{j1} \\ \vdots \\ x_{j7} \end{bmatrix}$$

so that, in matrix notation, the saturated model (6.13) is E($\mathbf{y}$)=$\mathbf{X}\boldsymbol{\beta}$ with

$$\mathbf{y} = \begin{bmatrix} \mathbf{y}_1 \\ \mathbf{y}_2 \\ \mathbf{y}_3 \end{bmatrix}, \quad \boldsymbol{\beta} = \begin{bmatrix} \mu_1 \\ \mu_2 \\ \mu_3 \\ \gamma \end{bmatrix} \quad \text{and} \quad \mathbf{X} = \begin{bmatrix} \mathbf{1} & \mathbf{0} & \mathbf{0} & \mathbf{x}_1 \\ \mathbf{0} & \mathbf{1} & \mathbf{0} & \mathbf{x}_2 \\ \mathbf{0} & \mathbf{0} & \mathbf{1} & \mathbf{x}_3 \end{bmatrix}$$

where $\mathbf{0}$ and $\mathbf{1}$ are vectors of length 7. Then

$$\mathbf{X}^T\mathbf{X} = \begin{bmatrix} 7 & 0 & 0 & 15 \\ 0 & 7 & 0 & 24 \\ 0 & 0 & 7 & 19 \\ 15 & 24 & 19 & 196 \end{bmatrix}, \quad \mathbf{X}^T\mathbf{y} = \begin{bmatrix} 31 \\ 53 \\ 47 \\ 398 \end{bmatrix}$$

and so

$$\mathbf{b} = \begin{bmatrix} 2.837 \\ 5.024 \\ 4.698 \\ 0.743 \end{bmatrix}.$$

Also $\mathbf{y}^T\mathbf{y} = 881$ and $\mathbf{b}^T\mathbf{X}^T\mathbf{y} = 870.698$ so for the saturated model (6.13)

$$\sigma^2 D_1 = \mathbf{y}^T\mathbf{y} - \mathbf{b}^T\mathbf{X}^T\mathbf{y} = 10.302.$$

For the reduced model (6.14)

$$\boldsymbol{\beta} = \begin{bmatrix} \mu \\ \gamma \end{bmatrix}, \quad \mathbf{X} = \begin{bmatrix} \mathbf{1} & \mathbf{x}_1 \\ \mathbf{1} & \mathbf{x}_2 \\ \mathbf{1} & \mathbf{x}_3 \end{bmatrix} \quad \text{so} \quad \mathbf{X}^T\mathbf{X} = \begin{bmatrix} 21 & 58 \\ 58 & 196 \end{bmatrix}$$

and

$$\mathbf{X}^T\mathbf{y} = \begin{bmatrix} 131 \\ 398 \end{bmatrix}.$$

Hence

$$\mathbf{b} = \begin{bmatrix} 3.447 \\ 1.011 \end{bmatrix}, \quad \mathbf{b}^T\mathbf{X}^T\mathbf{y} = 853.766 \quad \text{and so} \quad \sigma^2 D_0 = 27.234.$$

If we assume that the saturated model (6.13) is correct, then $D_1 \sim \chi^2(17)$. If the null hypothesis corresponding to model (6.14) is true then $D_0 \sim \chi^2(19)$ so

$$F = \frac{D_0 - D_1}{2\sigma^2} \bigg/ \frac{D_1}{17\sigma^2} \sim F(2, 17).$$

For these data

$$F = \frac{16.932}{2} \bigg/ \frac{10.302}{17} = 13.97$$

indicating a significant difference in achievement scores for the training methods, after adjustment for initial differences in aptitude. The usual presentation of this analysis is given in Table 6.13.

Table 6.13 *ANCOVA table for data in Table 6.11.*

Source of variation	Degrees of freedom	Sum of squares	Mean square	F
Mean and covariate	2	853.766		
Factor levels	2	16.932	8.466	13.97
Residuals	17	10.302	0.606	
Total	21	881.000		

6.6 General linear models

The term **general linear model** is used for Normal linear models with any combination of categorical and continuous explanatory variables. The factors may be **crossed**, as in Section 6.4.2., so that there are observations for each combination of levels of the factors. Alternatively, they may be **nested** as illustrated in the following example.

Table 6.14 shows a two-factor nested design which represents an experiment to compare two drugs (A_1 and A_2), one of which is tested in three hospitals (B_1, B_2 and B_3) and the other in two different hospitals (B_4 and B_5). We want to compare the effects of the two drugs and possible differences among hospitals using the same drug. In this case, the saturated model would be

$$\mathrm{E}(Y_{jkl}) = \mu + \alpha_1 + \alpha_2 + (\alpha\beta)_{11} + (\alpha\beta)_{12} + (\alpha\beta)_{13} + (\alpha\beta)_{24} + (\alpha\beta)_{25}$$

subject to some constraints (the corner point constraints are $\alpha_1 = 0, (\alpha\beta)_{11} = 0$ and $(\alpha\beta)_{24} = 0$). Hospitals B_1, B_2 and B_3 can only be compared *within* drug A_1 and hospitals B_4 and B_5 *within* A_2.

Table 6.14 *Nested two-factor experiment.*

		Drug A_1		Drug A_2	
Hospitals	B_1	B_2	B_3	B_4	B_5
Responses	Y_{111}	Y_{121}	Y_{131}	Y_{241}	Y_{251}
	$\vdots$	$\vdots$	$\vdots$	$\vdots$	$\vdots$
	Y_{11n_1}	Y_{12n_2}	Y_{13n_3}	Y_{24n_4}	Y_{25n_5}

Analysis for nested designs is not in principle, different from analysis for studies with crossed factors. Key assumptions for general linear models are that the response variable has the Normal distribution, the response and explanatory variables are *linearly* related and the variance σ^2 is the same for all responses. For the models considered in this chapter, the responses are also assumed to be independent (though this assumption is dropped in Chapter 11). All these assumptions can be examined through the use of residuals (Section 6.2.6). If they are not justified, for example, because the residuals have a skewed distribution, then it is usually worthwhile to consider transforming the response variable so that the assumption of Normality is more plausible. A useful tool, now available in many statistical programs, is the **Box-Cox transformation** (Box and Cox, 1964). Let y be the original variable and y^* the transformed one, then the function

$$y^* = \begin{cases} \dfrac{y^\lambda - 1}{\lambda} & , \quad \lambda \neq 0 \\ \log y & , \quad \lambda = 0 \end{cases}$$

provides a family of transformations. For example, except for a location shift, $\lambda = 1$ leaves y unchanged; $\lambda = \frac{1}{2}$ corresponds to taking the square root; $\lambda = -1$ corresponds to the reciprocal; and $\lambda = 0$ corresponds to the logarithmic transformation. The value of λ which produces the 'most Normal' distribution can be estimated by the method of maximum likelihood.

Similarly, transformation of continuous explanatory variables may improve the linearity of relationships with the response.

6.7 Exercises

6.1 Table 6.15 shows the average apparent per capita consumption of sugar (in kg per year) in Australia, as refined sugar and in manufactured foods (from Australian Bureau of Statistics, 1998).

Table 6.15 *Australian sugar consumption.*

Period	Refined sugar	Sugar in manufactured food
1936-39	32.0	16.3
1946-49	31.2	23.1
1956-59	27.0	23.6
1966-69	21.0	27.7
1976-79	14.9	34.6
1986-89	8.8	33.9

(a) Plot sugar consumption against time separately for refined sugar and sugar in manufactured foods. Fit simple linear regression models to summarize the pattern of consumption of each form of sugar. Calculate 95% confidence intervals for the average annual change in consumption for each form.

(b) Calculate the total average sugar consumption for each period and plot these data against time. Using suitable models test the hypothesis that total sugar consumption did not change over time.

6.2 Table 6.16 shows response of a grass and legume pasture system to various quantities of phosphorus fertilizer (data from D. F. Sinclair; the results were reported in Sinclair and Probert, 1986). The total yield, of grass and legume together, and amount of phosphorus (K) are both given in kilograms per hectare. Find a suitable model for describing the relationship between yield and quantity of fertilizer.

(a) Plot yield against phosphorus to obtain an approximately linear relationship – you may need to try several transformations of either or both variables in order to achieve approximate linearity.

(b) Use the results of (a) to specify a possible model. Fit the model.

(c) Calculate the standardized residuals for the model and use appropriate plots to check for any systematic effects that might suggest alternative models and to investigate the validity of any assumptions made.

6.3 Analyze the carbohydrate data in Table 6.3 using appropriate software (or, preferably, repeat the analyses using several different regression programs and compare the results).

Table 6.16 *Yield of grass and legume pasture and phosphorus levels (K).*

K	Yield	K	Yield	K	Yield
0	1753.9	15	3107.7	10	2400.0
40	4923.1	30	4415.4	5	2861.6
50	5246.2	50	4938.4	40	3723.0
5	3184.6	5	3046.2	30	4892.3
10	3538.5	0	2553.8	40	4784.6
30	4000.0	10	3323.1	20	3184.6
15	4184.6	40	4461.5	0	2723.1
40	4692.3	20	4215.4	50	4784.6
20	3600.0	40	4153.9	15	3169.3

(a) Plot the responses y against each of the explanatory variables x_1, x_2 and x_3 to see if y appears to be linearly related to them.

(b) Fit the model (6.6) and examine the residuals to assess the adequacy of the model and the assumptions.

(c) Fit the models

$$\mathrm{E}(Y_i) = \beta_0 + \beta_1 x_{i1} + \beta_3 x_{i3}$$

and

$$\mathrm{E}(Y_i) = \beta_0 + \beta_3 x_{i3},$$

(note the variable x_2, relative weight, is omitted from both models) and use these to test the hypothesis: $\beta_1 = 0$. Compare your results with Table 6.5.

6.4 It is well known that the concentration of cholesterol in blood serum increases with age but it is less clear whether cholesterol level is also associated with body weight. Table 6.17 shows for thirty women serum cholesterol (millimoles per liter), age (years) and body mass index (weight divided by height squared, where weight was measured in kilograms and height in meters). Use multiple regression to test whether serum cholesterol is associated with body mass index when age is already included in the model.

6.5 Table 6.18 shows plasma inorganic phosphate levels (mg/dl) one hour after a standard glucose tolerance test for obese subjects, with or without hyperinsulinemia, and controls (data from Jones, 1987).

(a) Perform a one-factor analysis of variance to test the hypothesis that there are no mean differences among the three groups. What conclusions can you draw?

(b) Obtain a 95% confidence interval for the difference in means between the two obese groups.

Table 6.17 *Cholesterol (CHOL), age and body mass index (BMI) for thirty women.*

CHOL	*Age*	*BMI*	*CHOL*	*Age*	*BMI*
5.94	52	20.7	6.48	65	26.3
4.71	46	21.3	8.83	76	22.7
5.86	51	25.4	5.10	47	21.5
6.52	44	22.7	5.81	43	20.7
6.80	70	23.9	4.65	30	18.9
5.23	33	24.3	6.82	58	23.9
4.97	21	22.2	6.28	78	24.3
8.78	63	26.2	5.15	49	23.8
5.13	56	23.3	2.92	36	19.6
6.74	54	29.2	9.27	67	24.3
5.95	44	22.7	5.57	42	22.0
5.83	71	21.9	4.92	29	22.5
5.74	39	22.4	6.72	33	24.1
4.92	58	20.2	5.57	42	22.7
6.69	58	24.4	6.25	66	27.3

Table 6.18 *Plasma phosphate levels in obese and control subjects.*

Hyperinsulinemic obese	Non-hyperinsulinemic obese	Controls
2.3	3.0	3.0
4.1	4.1	2.6
4.2	3.9	3.1
4.0	3.1	2.2
4.6	3.3	2.1
4.6	2.9	2.4
3.8	3.3	2.8
5.2	3.9	3.4
3.1		2.9
3.7		2.6
3.8		3.1
		3.2

(c) Using an appropriate model examine the standardized residuals for all the observations to look for any systematic effects and to check the Normality assumption.

6.6 The weights (in grams) of machine components of a standard size made by four different workers on two different days are shown in Table 6.19; five components were chosen randomly from the output of each worker on each

Table 6.19 *Weights of machine components made by workers on different days.*

	Workers			
	1	2	3	4
Day 1	35.7	38.4	34.9	37.1
	37.1	37.2	34.3	35.5
	36.7	38.1	34.5	36.5
	37.7	36.9	33.7	36.0
	35.3	37.2	36.2	33.8
Day 2	34.7	36.9	32.0	35.8
	35.2	38.5	35.2	32.9
	34.6	36.4	33.5	35.7
	36.4	37.8	32.9	38.0
	35.2	36.1	33.3	36.1

day. Perform an analysis of variance to test for differences among workers, among days, and possible interaction effects. What are your conclusions?

6.7 For the balanced data in Table 6.9, the analyses in Section 6.4.2 showed that the hypothesis tests were independent. An alternative specification of the design matrix for the saturated model (6.9) with the corner point constraints $\alpha_1 = \beta_1 = (\alpha\beta)_{11} = (\alpha\beta)_{12} = (\alpha\beta)_{21} = (\alpha\beta)_{31} = 0$ so that

$$\boldsymbol{\beta} = \begin{bmatrix} \mu \\ \alpha_2 \\ \alpha_3 \\ \beta_2 \\ (\alpha\beta)_{22} \\ (\alpha\beta)_{32} \end{bmatrix} \quad \text{is} \quad \mathbf{X} = \begin{bmatrix} 1 & -1 & -1 & -1 & 1 & 1 \\ 1 & -1 & -1 & -1 & 1 & 1 \\ 1 & -1 & -1 & 1 & -1 & -1 \\ 1 & -1 & -1 & 1 & -1 & -1 \\ 1 & 1 & 0 & -1 & -1 & 0 \\ 1 & 1 & 0 & -1 & -1 & 0 \\ 1 & 1 & 0 & 1 & 1 & 0 \\ 1 & 1 & 0 & 1 & 1 & 0 \\ 1 & 0 & 1 & -1 & 0 & -1 \\ 1 & 0 & 1 & -1 & 0 & -1 \\ 1 & 0 & 1 & 1 & 0 & 1 \\ 1 & 0 & 1 & 1 & 0 & 1 \end{bmatrix}$$

where the columns of $\mathbf{X}$ corresponding to the terms $(\alpha\beta)_{jk}$ are the products of columns corresponding to terms α_j and β_k.

(a) Show that $\mathbf{X}^T\mathbf{X}$ has the block diagonal form described in Section 6.2.5. Fit the model (6.9) and also models (6.10) to (6.12) and verify that the results in Table 6.9 are the same for this specification of $\mathbf{X}$.

(b) Show that the estimates for the mean of the subgroup with treatments A_3 and B_2 for two different models are the same as the values given at the end of Section 6.4.2.

6.8 Table 6.20 shows the data from a fictitious two-factor experiment.

(a) Test the hypothesis that there are no interaction effects.

(b) Test the hypothesis that there is no effect due to factor A

(i) by comparing the models

$$\mathrm{E}(Y_{jkl}) = \mu + \alpha_j + \beta_k \quad \text{and} \quad \mathrm{E}(Y_{jkl}) = \mu + \beta_k;$$

(ii) by comparing the models

$$\mathrm{E}(Y_{jkl}) = \mu + \alpha_j \quad \text{and} \quad \mathrm{E}(Y_{jkl}) = \mu.$$

Explain the results.

Table 6.20 *Two factor experiment with unbalanced data.*

	Factor B	
Factor A	B_1	B_2
A_1	5	3, 4
A_2	6, 4	4, 3
A_3	7	6, 8

7

Binary Variables and Logistic Regression

7.1 Probability distributions

In this chapter we consider generalized linear models in which the outcome variables are measured on a binary scale. For example, the responses may be alive or dead, or present or absent. 'Success' and 'failure' are used as generic terms of the two categories.

First, we define the **binary random variable**

$$Z = \begin{cases} 1 \text{ if the outcome is a success} \\ 0 \text{ if the outcome is a failure} \end{cases}$$

with probabilities $\Pr(Z = 1) = \pi$ and $\Pr(Z = 0) = 1 - \pi$. If there are n such random variables $Z_1, ..., Z_n$ which are independent with $\Pr(Z_j = 1) = \pi_j$, then their joint probability is

$$\prod_{j=1}^{n} \pi_j^{z_j} (1 - \pi_j)^{1-z_j} = \exp\left[\sum_{j=1}^{n} z_j \log\left(\frac{\pi_j}{1 - \pi_j}\right) + \sum_{j=1}^{n} \log(1 - \pi_j)\right] \qquad (7.1)$$

which is a member of the exponential family (see equation (3.3)).

Next, for the case where the π_j's are all equal, we can define

$$Y = \sum_{j=1}^{n} Z_j$$

so that Y is the number of successes in n 'trials'. The random variable Y has the distribution *binomial*(n, π):

$$\Pr(Y = y) = \binom{n}{y} \pi^y (1 - \pi)^{n-y} \quad , \quad y = 0, 1, ..., n \qquad (7.2)$$

Finally, we consider the general case of N independent random variables $Y_1, Y_2, ..., Y_N$ corresponding to the numbers of successes in N different subgroups or strata (Table 7.1). If $Y_i \sim binomial(n_i, \pi_i)$ the log-likelihood function is

$$l(\pi_1, \ldots, \pi_N; y_1, \ldots, y_N) = \left[\sum_{i=1}^{N} y_i \log\left(\frac{\pi_i}{1 - \pi_i}\right) + n_i \log(1 - \pi_i) + \log\binom{n_i}{y_i}\right]. \qquad (7.3)$$

Table 7.1 *Frequencies for N binomial distributions.*

	Subgroups			
	1	2	...	N
Successes	Y_1	Y_2	...	Y_N
Failures	$n_1 - Y_1$	$n_2 - Y_2$	...	$n_N - Y_N$
Totals	n_1	n_2	...	n_N

7.2 Generalized linear models

We want to describe the proportion of successes, $P_i = Y_i/n_i$, in each subgroup in terms of factor levels and other explanatory variables which characterize the subgroup. As $\mathrm{E}(Y_i) = n_i\pi_i$ and so $\mathrm{E}(P_i) = \pi_i$, we model the probabilities π_i as

$$g(\pi_i) = \mathbf{x}_i^T\boldsymbol{\beta}$$

where $\mathbf{x}_i$ is a vector of explanatory variables (dummy variables for factor levels and measured values for covariates), $\boldsymbol{\beta}$ is a vector of parameters and g is a link function.

The simplest case is the **linear model**

$$\pi = \mathbf{x}^T\boldsymbol{\beta}.$$

This is used in some practical applications but it has the disadvantage that although π is a probability, the fitted values $\mathbf{x}^T\mathbf{b}$ may be less than zero or greater than one.

To ensure that π is restricted to the interval [0,1] it is often modelled using a cumulative probability distribution

$$\pi = \int_{-\infty}^{t} f(s)ds$$

where $f(s) \geqslant 0$ and $\int_{-\infty}^{\infty} f(s)ds = 1$. The probability density function $f(s)$ is called the **tolerance distribution**. Some commonly used examples are considered in Section 7.3.

7.3 Dose response models

Historically, one of the first uses of regression-like models for binomial data was for bioassay results (Finney, 1973). Responses were the proportions or percentages of 'successes'; for example, the proportion of experimental animals killed by various dose levels of a toxic substance. Such data are sometimes called **quantal responses**. The aim is to describe the probability of 'success', π, as a function of the dose, x; for example, $g(\pi) = \beta_1 + \beta_2 x$.

If the tolerance distribution $f(s)$ is the uniform distribution on the interval

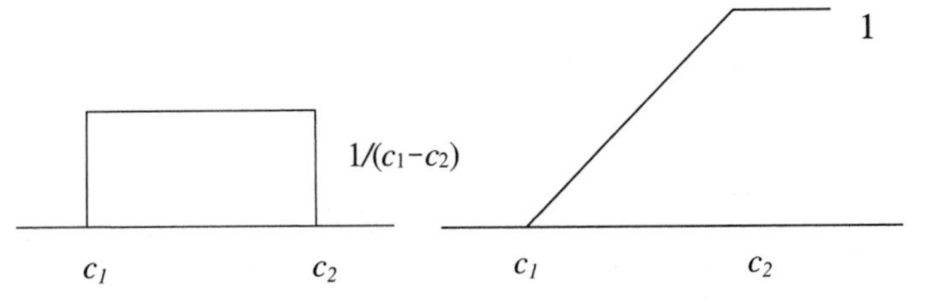

Figure 7.1 *Uniform distribution: $f(s)$ and π.*

$[c_1, c_2]$

$$f(s) = \begin{cases} \dfrac{1}{c_2 - c_1} & \text{if } c_1 \leqslant s \leqslant c_2 \\ 0 & \text{otherwise} \end{cases},$$

then

$$\pi = \int_{c_1}^{x} f(s)ds = \frac{x - c_1}{c_2 - c_1} \qquad \text{for } c_1 \leqslant x \leqslant c_2$$

(see Figure 7.1). This equation has the form $\pi = \beta_1 + \beta_2 x$ where

$$\beta_1 = \frac{-c_1}{c_2 - c_1} and \beta_2 = \frac{1}{c_2 - c_1}.$$

This **linear model** is equivalent to using the identity function as the link function g and imposing conditions on x, β_1 and β_2 corresponding to $c_1 \leq x \leq c_2$. These extra conditions mean that the standard methods for estimating β_1 and β_2 for generalized linear models cannot be directly applied. In practice, this model is not widely used.

One of the original models used for bioassay data is called the **probit model**. The Normal distribution is used as the tolerance distribution (see Figure 7.2).

$$\begin{aligned} \pi &= \frac{1}{\sigma\sqrt{2\pi}} \int_{-\infty}^{x} \exp\left[-\frac{1}{2}\left(\frac{s-\mu}{\sigma}\right)^2\right] ds \\ &= \Phi\left(\frac{x-\mu}{\sigma}\right) \end{aligned}$$

where Φ denotes the cumulative probability function for the standard Normal distribution $N(0, 1)$. Thus

$$\Phi^{-1}(\pi) = \beta_1 + \beta_2 x$$

where $\beta_1 = -\mu/\sigma$ and $\beta_2 = 1/\sigma$ and the link function g is the inverse cumulative Normal probability function Φ^{-1}. Probit models are used in several areas of biological and social sciences in which there are natural interpretations of the model; for example, $x = \mu$ is called the **median lethal dose** LD(50)

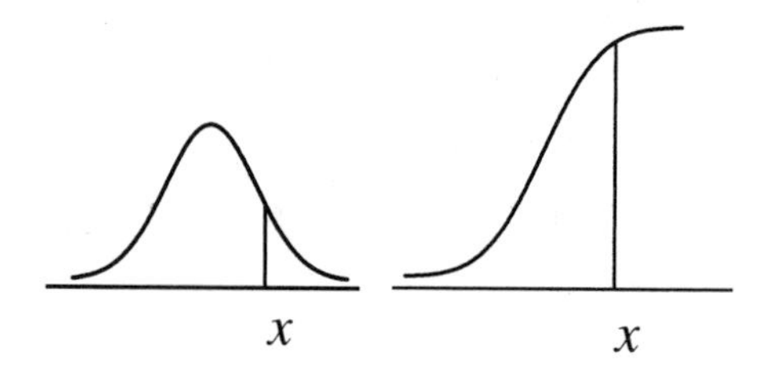

Figure 7.2 *Normal distribution: $f(s)$ and π.*

because it corresponds to the dose that can be expected to kill half of the animals.

Another model that gives numerical results very much like those from the probit model, but which computationally is somewhat easier, is the **logistic** or **logit model**. The tolerance distribution is

$$f(s) = \frac{\beta_2 \exp(\beta_1 + \beta_2 s)}{[1 + \exp(\beta_1 + \beta_2 s)]^2}$$

so

$$\pi = \int_{-\infty}^{x} f(s)ds = \frac{\exp(\beta_1 + \beta_2 x)}{1 + \exp(\beta_1 + \beta_2 x)}.$$

This gives the link function

$$\log\left(\frac{\pi}{1-\pi}\right) = \beta_1 + \beta_2 x.$$

The term $\log[\pi/(1-\pi)]$ is sometimes called the **logit function** and it has a natural interpretation as the logarithm of odds (see Exercise 7.2). The logistic model is widely used for binomial data and is implemented in many statistical programs. The shapes of the functions $f(s)$ and $\pi(x)$ are similar to those for the probit model (Figure 7.2) except in the tails of the distributions (see Cox and Snell, 1989).

Several other models are also used for dose response data. For example, if the **extreme value distribution**

$$f(s) = \beta_2 \exp\left[(\beta_1 + \beta_2 s) - \exp\left(\beta_1 + \beta_2 s\right)\right]$$

is used as the tolerance distribution then

$$\pi = 1 - \exp\left[-\exp\left(\beta_1 + \beta_2 x\right)\right]$$

and so $\log[-\log(1-\pi)] = \beta_1 + \beta_2 x$. This link, $\log[-\log(1-\pi)]$, is called the **complementary log log function**. The model is similar to the logistic and probit models for values of π near 0.5 but differs from them for π near 0 or 1. These models are illustrated in the following example.

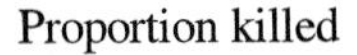

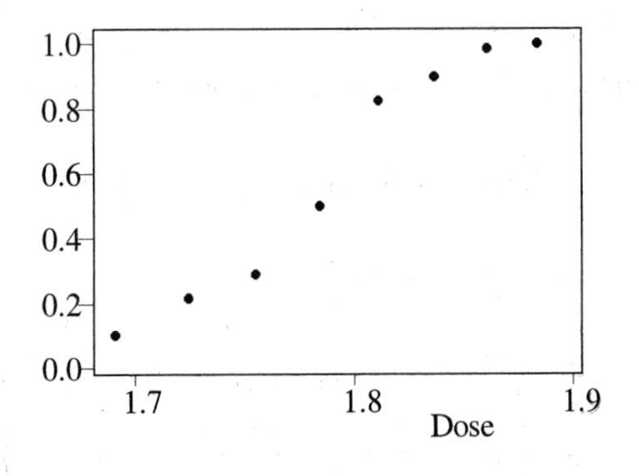

Figure 7.3 *Beetle mortality data from Table 7.2: proportion killed, $p_i = y_i/n_i$, plotted against dose, x_i ($\log_{10} CS_2 mgl^{-1}$).*

7.3.1 Example: Beetle mortality

Table 7.2 shows numbers of beetles dead after five hours exposure to gaseous carbon disulphide at various concentrations (data from Bliss, 1935). Figure

Table 7.2 *Beetle mortality data.*

Dose, x_i ($\log_{10}CS_2mgl^{-1}$)	Number of beetles, n_i	Number killed, y_i
1.6907	59	6
1.7242	60	13
1.7552	62	18
1.7842	56	28
1.8113	63	52
1.8369	59	53
1.8610	62	61
1.8839	60	60

7.3 shows the proportions $p_i = y_i/n_i$ plotted against dose x_i (actually x_i is the logarithm of the quantity of carbon disulphide). We begin by fitting the logistic model

$$\pi_i = \frac{\exp(\beta_1 + \beta_2 x_i)}{1 + \exp(\beta_1 + \beta_2 x_i)}$$

so

$$\log\left(\frac{\pi_i}{1 - \pi_i}\right) = \beta_1 + \beta_2 x_i$$

and

$$\log(1-\pi_i) = -\log\left[1+\exp\left(\beta_1+\beta_2 x_i\right)\right].$$

Therefore from equation (7.3) the log-likelihood function is

$$l = \sum_{i=1}^{N}\left[y_i\left(\beta_1+\beta_2 x_i\right) - n_i \log\left[1+\exp\left(\beta_1+\beta_2 x_i\right)\right] + \log\left(\begin{array}{c} n_i \\ y_i \end{array}\right)\right]$$

and the scores with respect to β_1 and β_2 are

$$\begin{aligned} U_1 &= \frac{\partial l}{\partial \beta_1} = \sum\left\{y_i - n_i\left[\frac{\exp\left(\beta_1+\beta_2 x_i\right)}{1+\exp\left(\beta_1+\beta_2 x_i\right)}\right]\right\} = \sum(y_i - n_i\pi_i) \\ U_2 &= \frac{\partial l}{\partial \beta_2} = \sum\left\{y_i x_i - n_i x_i\left[\frac{\exp\left(\beta_1+\beta_2 x_i\right)}{1+\exp\left(\beta_1+\beta_2 x_i\right)}\right]\right\} \\ &= \sum x_i(y_i - n_i\pi_i). \end{aligned}$$

Similarly the information matrix is

$$\mathfrak{I} = \left[\begin{array}{cc} \sum n_i\pi_i(1-\pi_i) & \sum n_i x_i \pi_i(1-\pi_i) \\ \sum n_i x_i\pi_i(1-\pi_i) & \sum n_i x_i^2\pi_i(1-\pi_i) \end{array}\right].$$

Maximum likelihood estimates are obtained by solving the iterative equation

$$\mathfrak{I}^{(m-1)}\mathbf{b}^{m} = \mathfrak{I}^{(m-1)}\mathbf{b}^{(m-1)} + \mathbf{U}^{(m-1)}$$

(from (4.22)) where the superscript (m) indicates the mth approximation and $\mathbf{b}$ is the vector of estimates. Starting with $b_1^{(0)} = 0$ and $b_2^{(0)} = 0$, successive approximations are shown in Table 7.3. The estimates converge by the sixth iteration. The table also shows the increase in values of the log-likelihood function (7.3), omitting the constant term $\log\left(\begin{array}{c} n_i \\ y_i \end{array}\right)$. The fitted values are $\widehat{y}_i = n_i\widehat{\pi}_i$ calculated at each stage (initially $\widehat{\pi}_i = \frac{1}{2}$ for all i).

For the final approximation, the estimated variance-covariance matrix for $\mathbf{b}$, $\left[\mathfrak{I}(\mathbf{b})^{-1}\right]$, is shown at the bottom of Table 7.3 together with the deviance

$$D = 2\sum_{i=1}^{N}\left[y_i \log\left(\frac{y_i}{\widehat{y}_i}\right) + (n_i - y_i)\log\left(\frac{n-y_i}{n-\widehat{y}_i}\right)\right]$$

(from Section 5.6.1).

The estimates and their standard errors are:

$$\begin{aligned} b_1 &= -60.72, \quad \text{standard error} = \sqrt{26.840} = 5.18 \\ \text{and} \quad b_2 &= 34.72, \quad \text{standard error} = \sqrt{8.481} = 2.91. \end{aligned}$$

If the model is a good fit of the data the deviance should approximately have the distribution $\chi^2(6)$ because there are $N = 8$ covariate patterns (i.e., different values of x_i) and $p = 2$ parameters. But the calculated value of D is almost twice the 'expected' value of 6 and is almost as large as the upper 5%

point of the $\chi^2(6)$ distribution, which is 12.59. This suggests that the model does not fit particularly well.

Table 7.3 *Fitting a linear logistic model to the beetle mortality data.*

		Initial estimate	Approximation First	Second	Sixth
β_1		0	-37.856	-53.853	-60.717
β_2		0	21.337	30.384	34.270
log-likelihood		-333.404	-200.010	-187.274	-186.235
Observations		**Fitted values**			
y_1	6	29.5	8.505	4.543	3.458
y_2	13	30.0	15.366	11.254	9.842
y_3	18	31.0	24.808	23.058	22.451
y_4	28	28.0	30.983	32.947	33.898
y_5	52	31.5	43.362	48.197	50.096
y_6	53	29.5	46.741	51.705	53.291
y_7	61	31.0	53.595	58.061	59.222
y_8	60	30.0	54.734	58.036	58.743

$$[\mathfrak{I}(\mathbf{b})]^{-1} = \begin{bmatrix} 26.840 & -15.082 \\ -15.082 & 8.481 \end{bmatrix}, \quad D = 11.23$$

Several alternative models were fitted to the data. The results are shown in Table 7.4. Among these models the extreme value model appears to fit the data best.

7.4 General logistic regression model

The simple linear logistic model $\log[\pi_i/(1-\pi_i)] = \beta_1 + \beta_2 x_i$ used in Example 7.3.1 is a special case of the general logistic regression model

$$\text{logit } \pi_i = \log\left(\frac{\pi_i}{1-\pi_i}\right) = \mathbf{x}_i^T\boldsymbol{\beta}$$

where $\mathbf{x}_i$ is a vector continuous measurements corresponding to covariates and dummy variables corresponding to factor levels and $\boldsymbol{\beta}$ is the parameter vector. This model is very widely used for analyzing data involving binary or binomial responses and several explanatory variables. It provides a powerful technique analogous to multiple regression and ANOVA for continuous responses.

Maximum likelihood estimates of the parameters $\boldsymbol{\beta}$, and consequently of the probabilities $\pi_i = g(\mathbf{x}_i^T\boldsymbol{\beta})$, are obtained by maximizing the log-likelihood

Table 7.4 *Comparison of observed numbers killed with fitted values obtained from various dose-response models for the beetle mortality data. Deviance statistics are also given.*

Observed value of Y	Logistic model	Probit model	Extreme value model
6	3.46	3.36	5.59
13	9.84	10.72	11.28
18	22.45	23.48	20.95
28	33.90	33.82	30.37
52	50.10	49.62	47.78
53	53.29	53.32	54.14
61	59.22	59.66	61.11
60	58.74	59.23	59.95
D	11.23	10.12	3.45

function

$$l(\boldsymbol{\pi}; \mathbf{y}) = \sum_{i=1}^{N} \left[y_i \log \pi_i + (n_i - y_i) \log(1 - \pi_i) + \log \binom{n_i}{y_i} \right] \tag{7.4}$$

using the methods described in Chapter 4.

The estimation process is essentially the same whether the data are grouped as frequencies for each **covariate pattern** (i.e., observations with the same values of all the explanatory variables) or each observation is coded 0 or 1 and its covariate pattern is listed separately. If the data can be grouped, the response Y_i, the number of 'successes' for covariate pattern i, may be modelled by the binomial distribution. If each observation has a different covariate pattern, then $n_i = 1$ and the response Y_i is binary.

The deviance, derived in Section 5.6.1, is

$$D = 2 \sum_{i=1}^{N} \left[y_i \log \left(\frac{y_i}{\widehat{y}_i} \right) + (n_i - y_i) \log \left(\frac{n_i - y_i}{n_i - \widehat{y}_i} \right) \right]. \tag{7.5}$$

This has the form

$$D = 2 \sum o \log \frac{o}{e}$$

where o denotes the observed frequencies y_i and $(n_i - y_i)$ from the cells of Table 7.1 and e denotes the corresponding estimated expected frequencies or fitted values $\widehat{y}_i = n_i \widehat{\pi}_i$ and $(n_i - \widehat{y}_i) = (n_i - n_i \widehat{\pi}_i)$. Summation is over all $2 \times N$ cells of the table.

Notice that D does not involve any nuisance parameters (like σ^2 for Normal response data), so goodness of fit can be assessed and hypotheses can be tested directly using the approximation

$$D \sim \chi^2(N-p)$$

where p is the number of parameters estimated and N the number of covariate patterns.

The estimation methods and sampling distributions used for inference depend on asymptotic results. For small studies or situations where there are few observations for each covariate pattern, the asymptotic results may be poor approximations. However software, such as StatXact and Log Xact, has been developed using 'exact' methods so that the methods described in this chapter can be used even when sample sizes are small.

7.4.1 Example: Embryogenic anthers

The data in Table 7.5, cited by Wood (1978), are taken from Sangwan-Norrell (1977). They are numbers y_{jk} of embryogenic anthers of the plant species *Datura innoxia* Mill. obtained when numbers n_{jk} of anthers were prepared under several different conditions. There is one qualitative factor with two levels, a treatment consisting of storage at 3°C for 48 hours or a control storage condition, and one continuous explanatory variable represented by three values of centrifuging force. We will compare the treatment and control effects on the proportions after adjustment (if necessary) for centrifuging force.

Table 7.5 *Embryogenic anther data.*

		Centrifuging force (g)		
Storage condition		40	150	350
Control	y_{1k}	55	52	57
	n_{1k}	102	99	108
Treatment	y_{2k}	55	50	50
	n_{2k}	76	81	90

The proportions $p_{jk} = y_{jk}/n_{jk}$ in the control and treatment groups are plotted against x_k, the logarithm of the centrifuging force, in Figure 7.4. The response proportions appear to be higher in the treatment group than in the control group and, at least for the treated group, the response decreases with centrifuging force.

We will compare three logistic models for π_{jk}, the probability of the anthers being embryogenic, where $j = 1$ for the control group and $j = 2$ for the treatment group and $x_1 = \log 40 = 3.689, x_2 = \log 150 = 5.011$ and $x_3 = \log 350 = 5.858$.

Model 1: $\text{logit}\ \pi_{jk} = \alpha_j + \beta_j x_k$ (i.e., different intercepts and slopes);
Model 2: $\text{logit}\ \pi_{jk} = \alpha_j + \beta x_k$ (i.e., different intercepts but the same slope);
Model 3: $\text{logit}\ \pi_{jk} = \alpha + \beta x_k$ (i.e., same intercept and slope).

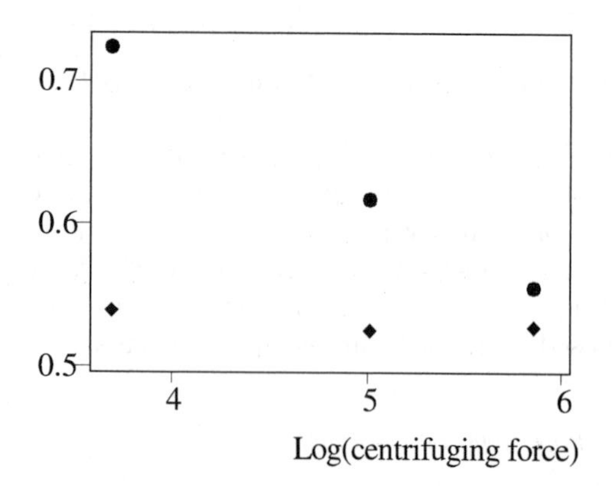

Figure 7.4 *Anther data from Table 7.5: proportion that germinated* $p_{jk} = y_{jk}/n_{jk}$ *plotted against* log *(centrifuging force); dots represent the treatment condition and diamonds represent the control condition.*

These models were fitted by the method of maximum likelihood. The results are summarized in Table 7.6.To test the null hypothesis that the slope is the same for the treatment and control groups, we use $D_2 - D_1 = 2.591$. From the tables for the $\chi^2(1)$ distribution, the significance level is between 0.1 and 0.2 and so we could conclude that the data provide little evidence against the null hypothesis of equal slopes. On the other hand, the power of this test is very low and both Figure 7.4 and the estimates for Model 1 suggest that although the slope for the control group may be zero, the slope for the treatment group is negative. Comparison of the deviances from Models 2 and 3 gives a test for equality of the control and treatment effects after a common adjustment for centrifuging force: $D_3 - D_2 = 0.491$, which is consistent with the hypothesis that the storage effects are not different.The observed proportions and the corresponding fitted values for Models 1 and 2 are shown in Table 7.7. Obviously, Model 1 fits the data very well but this is hardly surprising since four parameters have been used to describe six data points – such 'over-fitting' is not recommended!

7.5 Goodness of fit statistics

Instead of using maximum likelihood estimation we could estimate the parameters by minimizing the weighted sum of squares

$$S_w = \sum_{i=1}^{N} \frac{(y_i - n_i\pi_i)^2}{n_i\pi_i(1-\pi_i)}$$

since $\mathrm{E}(Y_i) = n_i\pi_i$ and $\mathrm{var}(Y_i) = n_i\pi_i(1-\pi_i)$.

Table 7.6 *Maximum likelihood estimates and deviances for logistic models for the embryogenic anther data (standard errors of estimates in brackets).*

Model 1	Model 2	Model 3
$a_1 = 0.234(0.628)$	$a_1 = 0.877(0.487)$	$a = 1.021(0.481)$
$a_2 - a_1 = 1.977(0.998)$	$a_2 - a_1 = 0.407(0.175)$	$b = -0.148(0.096)$
$b_1 = -0.023(0.127)$	$b = -0.155(0.097)$	
$b_2 - b_1 = -0.319(0.199)$		
$D_1 = 0.028$	$D_2 = 2.619$	$D_3 = 3.110$

Table 7.7 *Observed and expected frequencies for the embryogenic anther data for various models.*

Storage condition	Covariate value	Observed frequency	Expected frequencies Model 1	Model 2	Model 3
Control	x_1	55	54.82	58.75	62.91
	x_2	52	52.47	52.03	56.40
	x_3	57	56.72	53.22	58.18
Treatment	x_1	55	54.83	51.01	46.88
	x_2	50	50.43	50.59	46.14
	x_3	50	49.74	53.40	48.49

This is equivalent to minimizing the **Pearson chi-squared statistic**

$$X^2 = \sum \frac{(o-e)^2}{e}$$

where o represents the observed frequencies in Table 7.1, e represents the expected frequencies and summation is over all $2 \times N$ cells of the table. The reason is that

$$\begin{aligned} X^2 &= \sum_{i=1}^{N} \frac{(y_i - n_i\pi_i)^2}{n_i\pi_i} + \sum_{i=1}^{N} \frac{[(n_i - y_i) - n_i(1-\pi_i)]^2}{n_i(1-\pi_i)} \\ &= \sum_{i=1}^{N} \frac{(y_i - n_i\pi_i)^2}{n_i\pi_i(1-\pi_i)}(1 - \pi_i + \pi_i) = S_w. \end{aligned}$$

When X^2 is evaluated at the estimated expected frequencies, the statistic is

$$X^2 = \sum_{i=1}^{N} \frac{(y_i - n_i\widehat{\pi}_i)^2}{n_i\widehat{\pi}_i(1-\widehat{\pi}_i)} \tag{7.6}$$

which is asymptotically equivalent to the deviances in (7.5),

$$D = 2\sum_{i=1}^{N}\left[y_i \log\left(\frac{y_i}{n_i\widehat{\pi}_i}\right) + (n_i - y_i)\log\left(\frac{n_i - y_i}{n_i - n_i\widehat{\pi}_i}\right)\right].$$

The proof of the relationship between X^2 and D uses the Taylor series expansion of $s\log(s/t)$ about $s = t$, namely,

$$s\log\frac{s}{t} = (s-t) + \frac{1}{2}\frac{(s-t)^2}{t} + ... \quad .$$

Thus

$$\begin{aligned} D &= 2\sum_{i=1}^{N}\{(y_i - n_i\widehat{\pi}_i) + \frac{1}{2}\frac{(y_i - n_i\widehat{\pi}_i)^2}{n_i\widehat{\pi}_i} + [(n_i - y_i) - (n_i - n_i\widehat{\pi}_i)] \\ &\quad + \frac{1}{2}\frac{[(n_i - y_i) - (n_i - n_i\widehat{\pi}_i)]^2}{n_i - n_i\widehat{\pi}_i} + ...\} \\ &\cong \sum_{i=1}^{N}\frac{(y_i - n_i\widehat{\pi}_i)^2}{n_i\widehat{\pi}_i(1-\widehat{\pi}_i)} = X^2. \end{aligned}$$

The asymptotic distribution of D, under the hypothesis that the model is correct, is $D \sim \chi^2(N-p)$, therefore approximately $X^2 \sim \chi^2(N-p)$. The choice between D and X^2 depends on the adequacy of the approximation to the $\chi^2(N-p)$ distribution. There is some evidence to suggest that X^2 is often better than D because D is unduly influenced by very small frequencies (Cressie and Read, 1989). Both the approximations are likely to be poor, however, if the expected frequencies are too small (e.g., less than 1).

In particular, if each observation has a different covariate pattern so y_i is zero or one, then neither D nor X^2 provides a useful measure of fit. This can happen if the explanatory variables are continuous, for example. The most commonly used approach in this situation is due to Hosmer and Lemeshow (1980). Their idea was to group observations into categories on the basis of their predicted probabilities. Typically about 10 groups are used with approximately equal numbers of observations in each group. The observed numbers of successes and failures in each of the g groups are summarized as shown in Table 7.1. Then the Pearson chi-squared statistic for a $g \times 2$ contingency table is calculated and used as a measure of fit. We denote this **Hosmer-Lemeshow statistic** by X^2_{HL}. The sampling distribution of X^2_{HL} has been found by simulation to be approximately $\chi^2(g-2)$. The use of this statistic is illustrated in the example in Section 7.9.

Sometimes the log-likelihood function for the fitted model is compared with the log-likelihood function for a minimal model, in which the values π_i are all equal (in contrast to the saturated model which is used to define the deviance). Under the **minimal model** $\widetilde{\pi} = (\Sigma y_i) / (\Sigma n_i)$. Let $\widehat{\pi}_i$ denote the estimated probability for Y_i under the model of interest (so the fitted value is $\widehat{y}_i = n_i\widehat{\pi}_i$). The statistic is defined by

$$C = 2\left[l\left(\widehat{\boldsymbol{\pi}};\mathbf{y}\right) - l\left(\widetilde{\boldsymbol{\pi}};\mathbf{y}\right)\right]$$

where l is the log-likelihood function given by (7.4).

Thus

$$C = 2\sum\left[y_i \log\left(\frac{\widehat{y}_i}{n\widetilde{\pi}_i}\right) + (n_i - y_i)\log\left(\frac{n_i - \widehat{y}_i}{n_i - n_i\widetilde{\pi}_i}\right)\right]$$

From the results in Section 5.5, the approximate sampling distribution for C is $\chi^2(p-1)$ if all the p parameters except the intercept term β_1 are zero (see Exercise 7.4). Otherwise C will have a non-central distribution. Thus C is a test statistic for the hypothesis that none of the explanatory variables is needed for a parsimonious model. C is sometimes called the **likelihood ratio chi-squared statistic**.

In the beetle mortality example (Section 7.3.1), $C = 272.97$ with one degree of freedom, indicating that the slope parameter β_1 is definitely needed!

By analogy with R^2 for multiple linear regression (see Section 6.3.2) another statistic sometimes used is

$$\text{pseudo } R^2 = \frac{l(\widetilde{\boldsymbol{\pi}}; \mathbf{y}) - l(\widehat{\boldsymbol{\pi}}; \mathbf{y})}{l(\widetilde{\boldsymbol{\pi}}; \mathbf{y})}$$

which represents the proportional improvement in the log-likelihood function due to the terms in the model of interest, compared to the minimal model. This is produced by some statistical programs as a goodness of fit statistic.

7.6 Residuals

For logistic regression there are two main forms of residuals corresponding to the goodness of fit measures D and X^2. If there are m different covariate patterns then m residuals can be calculated. Let Y_k denote the number of successes, n_k the number of trials and $\widehat{\pi}_k$ the estimated probability of success for the kth covariate pattern.

The **Pearson, or chi-squared, residual** is

$$X_k = \frac{(y_k - n_k\widehat{\pi}_k)}{\sqrt{n_k\widehat{\pi}_k(1-\widehat{\pi}_k)}} \quad , k = 1, ..., m. \tag{7.7}$$

From (7.6), $\sum_{k=1}^{m} X_k^2 = X^2$, the Pearson chi-squared goodness of fit statistic. The **standardized Pearson residuals** are

$$r_{Pk} = \frac{X_k}{\sqrt{1-h_k}}$$

where h_k is the leverage, which is obtained from the hat matrix (see Section 6.2.6).

Deviance residuals can be defined similarly,

$$d_k = \text{sign}(y_k - n_k\widehat{\pi}_k)\left\{2\left[y_k \log\left(\frac{y_k}{n_k\widehat{\pi}_k}\right) + (n_k - y_k)\log\left(\frac{n_k - y_k}{n_k - n_k\widehat{\pi}_k}\right)\right]\right\}^{1/2} \tag{7.8}$$

where the term $\text{sign}(y_k - n_k\widehat{\pi}_k)$ ensures that d_k has the same sign as X_k.

From equation (7.5), $\sum_{k=1}^{m} d_k^2 = D$, the deviance. Also standardized deviance residuals are defined by

$$r_{Dk} = \frac{d_k}{\sqrt{1-h_k}}.$$

These residuals can be used for checking the adequacy of a model, as described in Section 2.3.4. For example, they should be plotted against each continuous explanatory variable in the model to check if the assumption of linearity is appropriate and against other possible explanatory variables not included in the model. They should be plotted in the order of the measurements, if applicable, to check for serial correlation. Normal probability plots can also be used because the standardized residuals should have, approximately, the standard Normal distribution $N(0,1)$, provided the numbers of observations for each covariate pattern are not too small.

If the data are binary, or if n_i is small for most covariate patterns, then there are few distinct values of the residuals and the plots may be relatively uninformative. In this case, it may be necessary to rely on the aggregated goodness of fit statistics X^2 and D and other diagnostics (see Section 7.7).

For more details about the use of residuals for binomial and binary data see Chapter 5 of Collett (1991), for example.

7.7 Other diagnostics

By analogy with the statistics used to detect influential observations in multiple linear regression, the statistics delta-beta, delta-chi-squared and delta-deviance are also available for logistic regression (see Section 6.2.7).

For binary or binomial data there are additional issues to consider. The first is to check the choice of the link function. Brown (1982) developed a test for the logit link which is implemented in some software. The approach suggested by Aranda-Ordaz (1981) is to consider a more general family of link functions

$$g(\pi, \alpha) = \log\left[\frac{(1-\pi)^{-\alpha} - 1}{\alpha}\right].$$

If $\alpha = 1$ then $g(\pi) = \log\left[\pi/(1-\pi)\right]$, the logit link. As $\alpha \to 0$, then $g(\pi) \to \log\left[-\log(1-\pi)\right]$, the complementary log-log link. In principle, an optimal value of α can be estimated from the data, but the process requires several steps. In the absence of suitable software to identify the best link function it is advisable to experiment with several alternative links.

The second issue in assessing the adequacy of models for binary or binomial data is **overdispersion**. Observations Y_i which might be expected to correspond to the binomial distribution may have variance greater than $n_i\pi_i(1-\pi_i)$. There is an indicator of this problem if the deviance D is much greater than the expected value of $N-p$. This could be due to inadequate specification of the model (e.g., relevant explanatory variables have been omitted or the link function is incorrect) or to a more complex structure. One approach is to include an extra parameter ϕ in the model so that $\text{var}(Y_i) = n_i\pi_i(1-\pi_i)\phi$.

This is implemented in various ways in statistical software. Another possible explanation for overdispersion is that the Y_i's are not independent. Methods for modelling correlated data are outlined in Chapter 11. For a detailed discussion of overdispersion for binomial data, see Collett (1991), Chapter 6.

7.8 Example: Senility and WAIS

A sample of elderly people was given a psychiatric examination to determine whether symptoms of senility were present. Other measurements taken at the same time included the score on a subset of the Wechsler Adult Intelligent Scale (WAIS). The data are shown in Table 7.8.

Table 7.8 *Symptoms of senility (s=1 if symptoms are present and s=0 otherwise) and WAIS scores (x) for N=54 people.*

x	s	x	s	x	s	x	s	x	s
9	1	7	1	7	0	17	0	13	0
13	1	5	1	16	0	14	0	13	0
6	1	14	1	9	0	19	0	9	0
8	1	13	0	9	0	9	0	15	0
10	1	16	0	11	0	11	0	10	0
4	1	10	0	13	0	14	0	11	0
14	1	12	0	15	0	10	0	12	0
8	1	11	0	13	0	16	0	4	0
11	1	14	0	10	0	10	0	14	0
7	1	15	0	11	0	16	0	20	0
9	1	18	0	6	0	14	0		

The data in Table 7.8 are binary although some people have the same WAIS scores and so there are $m = 17$ different covariate patterns (see Table 7.9). Let Y_i denote the number of people with symptoms among n_i people with the ith covariate pattern. The logistic regression model

$$\log\left(\frac{\pi_i}{1-\pi_i}\right) = \beta_1 + \beta_2 x_i; \quad Y_i \sim binomial(n_i, \pi_i) \quad i = 1, \ldots, m,$$

was fitted with the following results:

$$\begin{aligned}
&b_1 = 2.404, \text{ standard error } (b_1) = 1.192, \\
&b_2 = -0.3235, \text{ standard error } (b_2) = 0.1140, \\
&X^2 = \sum X_i^2 = 8.083 \text{ and } D = \sum d_i^2 = 9.419.
\end{aligned}$$

As there are $m = 17$ covariate patterns and $p = 2$ parameters, X^2 and D can be compared with $\chi^2(15)$ – by these criteria the model appears to fit well.

For the minimal model, without x, the maximum value of the log-likelihood function is $l(\widetilde{\pi}, \mathbf{y}) = -30.9032$. For the model with x, the corresponding value is $l(\widehat{\boldsymbol{\pi}}, \mathbf{y}) = -25.5087$. Therefore, from Section 7.5, $C = 10.789$ which is highly

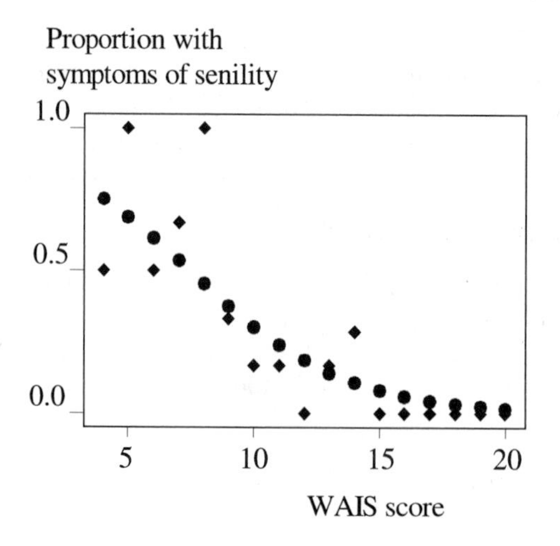

Figure 7.5 *Relationship between presence of symptoms and WAIS score from data in Tables 7.8 and 7.9; dots represent estimated probabilities and diamonds represent observed proportions.*

significant compared with $\chi^2(1)$, showing that the slope parameter is non-zero. Also pseudo $R^2 = 0.17$ which suggests the model is not particularly good.

Figure 7.5 shows the observed relative frequencies y_i/n_i for each covariate pattern and the fitted probabilities $\widehat{\pi}_i$ plotted against WAIS score, x (for $i = 1, ..., m$). The model appears to fit better for higher values of x.

Table 7.9 shows the covariate patterns, estimates $\widehat{\pi}_i$ and the corresponding chi-squared and deviance residuals calculated using equations (7.7) and (7.8) respectively.

The residuals and associated residual plots (not shown) do not suggest that there are any unusual observations but the small numbers of observations for each covariate value make the residuals difficult to assess. The Hosmer Lemeshow approach provides some simplification; Table 7.10 shows the data in categories defined by grouping values of $\widehat{\pi}_i$ so that the total numbers of observations per category are approximately equal. For this illustration, $g = 3$ categories were chosen. The expected frequencies are obtained from the values in Table 7.9; there are $\Sigma n_i \widehat{\pi}_i$ with symptoms and $\Sigma n_i (1 - \widehat{\pi}_i)$ without symptoms for each category. The Hosmer Lemeshow statistic X^2_{HL} is obtained by calculating $X^2 = \Sigma \left[(o - e)^2/e\right]$ where the observed frequencies, o, and expected frequencies, e, are given in Table 7.10 and summation is over all 6 cells of the table; $X^2_{HL} = 1.15$ which is not significant when compared with the $\chi^2(1)$ distribution.

Table 7.9 *Covariate patterns and responses, estimated probabilities* ($\widehat{\pi}$)*, Pearson residuals* (X) *and deviances* (d) *for senility and WAIS.*

x	y	n	$\widehat{\pi}$	X	d
4	1	2	0.751	-0.826	-0.766
5	0	1	0.687	0.675	0.866
6	1	2	0.614	-0.330	-0.326
7	1	3	0.535	0.458	0.464
8	0	2	0.454	1.551	1.777
9	4	6	0.376	-0.214	-0.216
10	5	6	0.303	-0.728	-0.771
11	5	6	0.240	-0.419	-0.436
12	2	2	0.186	-0.675	-0.906
13	5	6	0.142	0.176	0.172
14	5	7	0.107	1.535	1.306
15	3	3	0.080	-0.509	-0.705
16	4	4	0.059	-0.500	-0.696
17	1	1	0.043	-0.213	-0.297
18	1	1	0.032	-0.181	-0.254
19	1	1	0.023	-0.154	-0.216
20	1	1	0.017	-0.131	-0.184
Sum	40	54			
			Sum of squares	8.084*	9.418*

* Sums of squares differ slightly from the goodness of fit statistics X^2 and D mentioned in the text due to rounding errors.

7.9 Exercises

7.1 The number of deaths from leukemia and other cancers among survivors of the Hiroshima atom bomb are shown in Table 7.11, classified by the radiation dose received. The data refer to deaths during the period 1950-59 among survivors who were aged 25 to 64 years in 1950 (from data set 13 of Cox and Snell, 1981, attributed to Otake, 1979). Obtain a suitable model to describe the dose-response relationship between radiation and the proportional mortality rates for leukemia.

7.2 **Odds ratios**. Consider a 2×2 contingency table from a prospective study in which people who were or were not exposed to some pollutant are followed up and, after several years, categorized according to the presence or absence of a disease. Table 7.12 shows the probabilities for each cell. The odds of disease for either exposure group is $O_i = \pi_i/(1-\pi_i)$, for $i = 1, 2$, and so the odds ratio

$$\phi = \frac{O_1}{O_2} = \frac{\pi_1(1-\pi_2)}{\pi_2(1-\pi_1)}$$

Table 7.10 *Hosmer-Lemeshow test for data in Table 7.9: observed frequencies (o) and expected frequencies (e) for numbers of people with or without symptoms, grouped by values of* $\widehat{\pi}$.

Values of $\widehat{\pi}$		≤ 0.107	$0.108 - 0.303$	> 0.303
Corresponding values of x		$14 - 20$	$10 - 13$	$4 - 9$
Number of people	o	2	3	9
with symptoms	e	1.335	4.479	8.186
Number of people	o	16	17	7
without symptoms	e	16.665	15.521	7.814
Total number of people		18	20	16

Table 7.11 *Deaths from leukemia and other cancers classified by radiation dose received from the Hiroshima atomic bomb.*

	Radiation dose (rads)					
Deaths	0	1-9	10-49	50-99	100-199	200+
Leukemia	13	5	5	3	4	18
Other cancers	378	200	151	47	31	33
Total cancers	391	205	156	50	35	51

is a measure of the relative likelihood of disease for the exposed and not exposed groups.

Table 7.12 *2×2 table for a prospective study of exposure and disease outcome.*

	Diseased	Not diseased
Exposed	π_1	$1-\pi_1$
Not exposed	π_2	$1-\pi_2$

(a) For the simple logistic model $\pi_i = e^{\beta_i}/(1+e^{\beta_i})$, show that if there is no difference between the exposed and not exposed groups (i.e., $\beta_1 = \beta_2$) then $\phi = 1$.

(b) Consider J 2×2 tables like Table 7.12, one for each level x_j of a factor, such as age group, with $j = 1, ..., J$. For the logistic model

$$\pi_{ij} = \frac{\exp(\alpha_i + \beta_i x_j)}{1+\exp(\alpha_i + \beta_i x_j)}, \qquad i = 1, 2, \quad j = 1, ..., J.$$

Show that $\log \phi$ is constant over all tables if $\beta_1 = \beta_2$ (McKinlay, 1978).

7.3 Tables 7.13 and 7.14 show the survival 50 years after graduation of men and women who graduated each year from 1938 to 1947 from various Faculties of the University of Adelaide (data compiled by J.A. Keats). The columns labelled S contain the number of graduates who survived and the columns labelled T contain the total number of graduates. There were insufficient women graduates from the Faculties of Medicine and Engineering to warrant analysis.

Table 7.13 *Fifty years survival for men after graduation from the University of Adelaide.*

Year of graduation	Faculty							
	Medicine		Arts		Science		Engineering	
	S	T	S	T	S	T	S	T
1938	18	22	16	30	9	14	10	16
1939	16	23	13	22	9	12	7	11
1940	7	17	11	25	12	19	12	15
1941	12	25	12	14	12	15	8	9
1942	24	50	8	12	20	28	5	7
1943	16	21	11	20	16	21	1	2
1944	22	32	4	10	25	31	16	22
1945	12	14	4	12	32	38	19	25
1946	22	34			4	5		
1947	28	37	13	23	25	31	25	35
Total	177	275	92	168	164	214	100	139

Table 7.14 *Fifty years survival for women after graduation from the University of Adelaide.*

Year of graduation	Faculty			
	Arts		Science	
	S	T	S	T
1938	14	19	1	1
1939	11	16	4	4
1940	15	18	6	7
1941	15	21	3	3
1942	8	9	4	4
1943	13	13	8	9
1944	18	22	5	5
1945	18	22	16	17
1946	1	1	1	1
1947	13	16	10	10
Total	126	157	58	61

(a) Are the proportions of graduates who survived for 50 years after graduation the same all years of graduation?

(b) Are the proportions of male graduates who survived for 50 years after graduation the same for all Faculties?

(c) Are the proportions of female graduates who survived for 50 years after graduation the same for Arts and Science?

(d) Is the difference between men and women in the proportion of graduates who survived for 50 years after graduation the same for Arts and Science?

7.4 Let $l(\mathbf{b}_{\min})$ denote the maximum value of the log-likelihood function for the minimal model with linear predictor $\mathbf{x}^T\boldsymbol{\beta} = \beta_1$ and let $l(\mathbf{b})$ be the corresponding value for a more general model $\mathbf{x}^T\boldsymbol{\beta} = \beta_1 + \beta_2 x_1 + ... + \beta_p x_{p-1}$.

(a) Show that the likelihood ratio chi-squared statistic is

$$C = 2\left[l(b) - l(b_{\min})\right] = D_0 - D_1$$

where D_0 is the deviance for the minimal model and D_1 is the deviance for the more general model.

(b) Deduce that if $\beta_2 = ... = \beta_p = 0$ then C has the central chi-squared distribution with $(p-1)$ degrees of freedom.

8

Nominal and Ordinal Logistic Regression

8.1 Introduction

If the response variable is categorical, with more then two categories, then there are two options for generalized linear models. One relies on generalizations of logistic regression from dichotomous responses, described in Chapter 7, to nominal or ordinal responses with more than two categories. This first approach is the subject of this chapter. The other option is to model the frequencies or counts for the covariate patterns as the response variables with Poisson distributions. The second approach, called **log-linear modelling**, is covered in Chapter 9.

For nominal or ordinal logistic regression one of the measured or observed categorical variables is regarded as the response, and all other variables are explanatory variables. For log-linear models, all the variables are treated alike. The choice of which approach to use in a particular situation depends on whether one variable is clearly a 'response' (for example, the outcome of a prospective study) or several variables have the same status (as may be the situation in a cross-sectional study). Additionally, the choice may depend on how the results are to be presented and interpreted. Nominal and ordinal logistic regression yield odds ratio estimates which are relatively easy to interpret if there are no interactions (or only fairly simple interactions). Log-linear models are good for testing hypotheses about complex interactions, but the parameter estimates are less easily interpreted.

This chapter begins with the multinomial distribution which provides the basis for modelling categorical data with more than two categories. Then the various formulations for nominal and ordinal logistic regression models are discussed, including the interpretation of parameter estimates and methods for checking the adequacy of a model. A numerical example is used to illustrate the methods.

8.2 Multinomial distribution

Consider a random variable Y with J categories. Let $\pi_1, \pi_2, ..., \pi_J$ denote the respective probabilities, with $\pi_1 + \pi_2 + ... + \pi_J = 1$. If there are n independent observations of Y which result in y_1 outcomes in category 1, y_2 outcomes in category 2, and so on, then let

$$\mathbf{y} = \begin{bmatrix} y_1 \\ y_2 \\ \vdots \\ y_J \end{bmatrix}, \quad \text{with} \sum_{j=1}^{J} y_j = n.$$

The multinomial distribution is

$$f(\mathbf{y}\,|n) = \frac{n!}{y_1!y_2!...y_J!}\pi_1^{y_1}\pi_2^{y_2}...\pi_J^{y_J}. \tag{8.1}$$

If $J = 2$, then $\pi_2 = 1 - \pi_1$, $y_2 = n - y_1$ and (8.1) is the binomial distribution; see (7.2). In general, (8.1) does not satisfy the requirements for being a member of the exponential family of distributions (3.3). However the following relationship with the Poisson distribution ensures that generalized linear modelling is appropriate.

Let $Y_1, ..., Y_J$ denote independent random variables with distributions $Y_j \sim Poisson(\lambda_j)$. Their joint probability distribution is

$$f(\mathbf{y}) = \prod_{j=1}^{J} \frac{\lambda_j^{y_j} e^{-\lambda_j}}{y_j!} \tag{8.2}$$

where

$$\mathbf{y} = \begin{bmatrix} y_1 \\ \vdots \\ y_J \end{bmatrix}.$$

Let $n = Y_1 + Y_2 + ... + Y_J$, then n is a random variable with the distribution $n \sim Poisson(\lambda_1 + \lambda_2 + ... + \lambda_J)$ (see, for example, Kalbfleisch, 1985, page 142). Therefore the distribution of $\mathbf{y}$ conditional on n is

$$f(\mathbf{y}\,|n) = \left[\prod_{j=1}^{J} \frac{\lambda_j^{y_j} e^{-\lambda_j}}{y_j!}\right] \Big/ \frac{(\lambda_1 + ... + \lambda_J)^n e^{-(\lambda_1 + ... + \lambda_J)}}{n!}$$

which can be simplified to

$$f(\mathbf{y}\,|n) = \left(\frac{\lambda_1}{\sum \lambda_k}\right)^{y_1} ... \left(\frac{\lambda_J}{\sum \lambda_k}\right)^{y_J} \frac{n!}{y_1!...y_J!}. \tag{8.3}$$

If $\pi_j = \lambda_j \left(\sum_{k=1}^{K} \lambda_k\right)$, for $j = 1, ..., J$, then (8.3) is the same as (8.1) and $\sum_{j=1}^{J} \pi_j = 1$, as required. Therefore the multinomial distribution can be regarded as the joint distribution of Poisson random variables, conditional upon their sum n. This result provides a justification for the use of generalized linear modelling.

For the multinomial distribution (8.1) it can be shown that $\mathrm{E}(Y_j) = n\pi_j$, $\mathrm{var}(Y_j) = n\pi_j(1 - \pi_j)$ and $\mathrm{cov}(Y_j, Y_k) = -n\pi_j\pi_k$ (see, for example, Agresti, 1990, page 44).

In this chapter models based on the binomial distribution are considered, because pairs of response categories are compared, rather than all J categories simultaneously.

8.3 Nominal logistic regression

Nominal logistic regression models are used when there is no natural order among the response categories. One category is arbitrarily chosen as the **reference category**. Suppose this is the first category. Then the logits for the other categories are defined by

$$\text{logit}(\pi_j) = \log\left(\frac{\pi_j}{\pi_1}\right) = \mathbf{x}_j^T \boldsymbol{\beta}_j, \quad \text{for } j = 2, ..., J. \tag{8.4}$$

The $(J-1)$ logit equations are used simultaneously to estimate the parameters $\boldsymbol{\beta}_j$. Once the parameter estimates $\mathbf{b}_j$ have been obtained, the linear predictors $\mathbf{x}_j^T \mathbf{b}_j$ can be calculated. From (8.4)

$$\widehat{\pi}_j = \widehat{\pi}_1 \exp\left(\mathbf{x}_j^T \mathbf{b}_j\right) \quad \text{for } j = 2, ..., J.$$

But $\widehat{\pi}_1 + \widehat{\pi}_2 + ... + \widehat{\pi}_J = 1$ so

$$\widehat{\pi}_1 = \frac{1}{1 + \sum_{j=2}^{J} \exp\left(\mathbf{x}_j^T \mathbf{b}_j\right)}$$

and

$$\widehat{\pi}_j = \frac{\exp\left(\mathbf{x}_j^T \mathbf{b}_j\right)}{1 + \sum_{j=2}^{J} \exp\left(\mathbf{x}_j^T \mathbf{b}_j\right)}, \quad \text{for } j = 2, ..., J.$$

Fitted values, or 'expected frequencies', for each covariate pattern can be calculated by multiplying the estimated probabilities $\widehat{\pi}_j$ by the total frequency of the covariate pattern.

The **Pearson chi-squared residuals** are given by

$$r_i = \frac{o_i - e_i}{\sqrt{e_i}} \tag{8.5}$$

where o_i and e_i are the observed and expected frequencies for $i = 1, ..., N$ where N is J times the number of distinct covariate patterns. The residuals can be used to assess the adequacy of the model.

Summary statistics for goodness of fit are analogous to those for binomial logistic regression:

(i) **Chi-squared statistic**

$$X^2 = \sum_{i=1}^{N} r_i^2; \tag{8.6}$$

(ii) **Deviance**, defined in terms of the maximum values of the log-likelihood function for the fitted model, $l(\mathbf{b})$, and for the maximal model, $l(\mathbf{b}_{\max})$,

$$D = 2\left[l(\mathbf{b}_{\max}) - l(\mathbf{b})\right]; \tag{8.7}$$

(iii) **Likelihood ratio chi-squared statistic**, defined in terms of the maximum value of the log likelihood function for the minimal model, $l(\mathbf{b}_{\min})$, and $l(\mathbf{b})$,

$$C = 2\left[l(\mathbf{b}) - l(\mathbf{b}_{\min})\right]; \tag{8.8}$$

(iv)

$$\text{Pseudo } R^2 = \frac{l(\mathbf{b}_{\min}) - l(\mathbf{b})}{l(\mathbf{b}_{\min})}. \tag{8.9}$$

If the model fits well then both X^2 and D have, asymptotically, the distribution $\chi^2(N-p)$ where p is the number of parameters estimated. C has the asymptotic distribution $\chi^2\left[p-(J-1)\right]$ because the minimal model will have one parameter for each logit defined in (8.4).

Often it is easier to interpret the effects of explanatory factors in terms of odds ratios than the parameters $\boldsymbol{\beta}$. For simplicity, consider a response variable with J categories and a binary explanatory variable x which denotes whether an 'exposure' factor is present ($x = 1$) or absent ($x = 0$). The odds ratio for exposure for response j ($j = 2, ..., J$) relative to the reference category $j = 1$ is

$$OR_j = \frac{\pi_{jp}}{\pi_{ja}} \Big/ \frac{\pi_{1p}}{\pi_{1a}}$$

where π_{jp} and π_{ja} denote the probabilities of response category j ($j = 1, ..., J$) according to whether exposure is present or absent, respectively. For the model

$$\log\left(\frac{\pi_j}{\pi_1}\right) = \beta_{0j} + \beta_{1j}x, \quad j = 2, ..., J$$

the log odds are

$$\begin{aligned}
\log\left(\frac{\pi_{ja}}{\pi_{1a}}\right) &= \beta_{0j} \quad \text{when } x = 0, \text{ indicating the exposure is absent, and} \\
\log\left(\frac{\pi_{jp}}{\pi_{1p}}\right) &= \beta_{0j} + \beta_{1j} \text{ when } x = 1, \text{ indicating the exposure is present.}
\end{aligned}$$

Therefore the logarithm of the odds ratio can be written as

$$\begin{aligned}
\log OR_j &= \log\left(\frac{\pi_{jp}}{\pi_{1p}}\right) - \log\left(\frac{\pi_{ja}}{\pi_{1a}}\right) \\
&= \beta_{1j}
\end{aligned}$$

Hence $OR_j = \exp(\beta_{1j})$ which is estimated by $\exp(b_{1j})$. If $\beta_{1j} = 0$ then $OR_j = 1$ which corresponds to the exposure factor having no effect. Also, for example, 95% confidence limits for OR_j are given by $\exp[b_{1j} \pm 1.96 \times \text{s.e.}(b_{1j})]$ where s.e.(b_{1j}) denotes the standard error of b_{1j}. Confidence intervals which do not include unity correspond to β values significantly different from zero.

For nominal logistic regression, the explanatory variables may be categorical or continuous. The choice of the reference category for the response variable will affect the parameter estimates $\mathbf{b}$ but not the estimated probabilities $\widehat{\boldsymbol{\pi}}$ or the fitted values.

The following example illustrates the main characteristic of nominal logistic regression.

8.3.1 *Example: Car preferences*

In a study of motor vehicle safety, men and women driving small, medium sized and large cars were interviewed about vehicle safety and their preferences for cars, and various measurements were made of how close they sat to the steering wheel (McFadden et al., 2000). There were 50 subjects in each of the six categories (two sexes and three car sizes). They were asked to rate how important various features were to them when they were buying a car. Table 8.1 shows the ratings for air conditioning and power steering, according to the sex and age of the subject (the categories 'not important' and 'of little importance' have been combined).

Table 8.1 *Importance of air conditioning and power steering in cars (row percentages in brackets*)*

		Response			
Sex	Age	No or little importance	Important	Very important	Total
Women	18-23	26 (58%)	12 (27%)	7 (16%)	45
	24-40	9 (20%)	21 (47%)	15 (33%)	45
	> 40	5 (8%)	14 (23%)	41 (68%)	60
Men	18-30	40 (62%)	17 (26%)	8 (12%)	65
	24-40	17 (39%)	15 (34%)	12 (27%)	44
	> 40	8 (20%)	15 (37%)	18 (44%)	41
Total		105	94	101	300

* row percentages may not add to 100 due to rounding.

The proportions of responses in each category by age and sex are shown in Figure 8.1. For these data the response, importance of air conditioning and power steering, is rated on an ordinal scale but for the purpose of this example the order is ignored and the 3-point scale is treated as nominal. The category 'no or little' importance is chosen as the reference category. Age is also ordinal, but initially we will regard it as nominal.

Table 8.2 shows the results of fitting the nominal logistic regression model with reference categories of 'Women' and '18-23 years', and

$$\log\left(\frac{\pi_j}{\pi_1}\right) = \beta_{0j} + \beta_{1j}x_1 + \beta_{2j}x_2 + \beta_{3j}x_3, \qquad j = 2, 3 \tag{8.10}$$

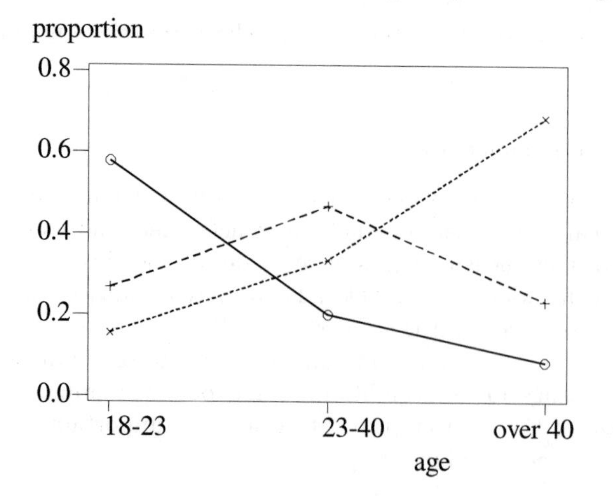

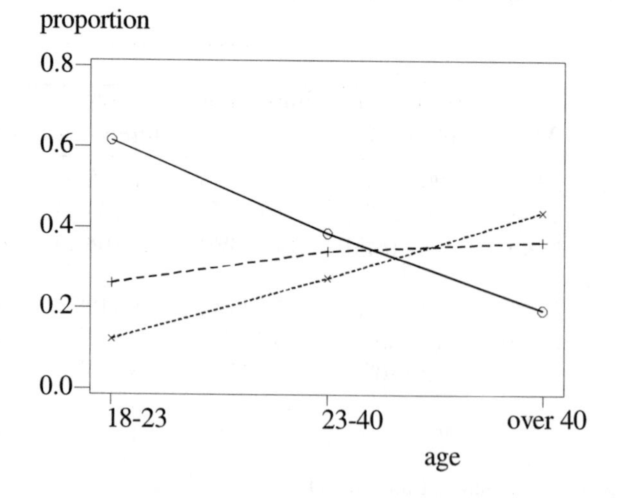

Figure 8.1 *Preferences for air conditioning and power steering: proportions of responses in each category by age and sex of respondents (solid lines denote 'no/little importance', dashed lines denote 'important' and dotted lines denote 'very important').*

where

$$x_1 = \begin{cases} 1 & \text{for men} \\ 0 & \text{for women} \end{cases}, \quad x_2 = \begin{cases} 1 & \text{for age 24-40 years} \\ 0 & \text{otherwise} \end{cases}$$

$$\text{and } x_3 = \begin{cases} 1 & \text{for age} > 40 \text{ years} \\ 0 & \text{otherwise} \end{cases}.$$

Table 8.2 *Results of fitting the nominal logistic regression model (8.10) to the data in Table 8.1.*

Parameter β	Estimate b (std. error)	Odds ratio, $OR = e^b$ (95% confidence interval)	
$\log(\pi_2/\pi_1)$: important vs. no/little importance			
β_{02}: constant	-0.591 (0.284)		
β_{12}: men	-0.388 (0.301)	0.68	(0.38, 1.22)
β_{22}: 24-40	1.128 (0.342)	3.09	(1.58, 6.04)
β_{32}: >40	1.588 (0.403)	4.89	(2.22, 10.78)
$\log(\pi_3/\pi_1)$: very important vs. no/little importance			
β_{03}: constant	-1.039 (0.331)		
β_{13}: men	-0.813 (0.321)	0.44	(0.24, 0.83)
β_{23}: 24-40	1.478 (0.401)	4.38	(2.00, 9.62)
β_{33}: > 40	2.917 (0.423)	18.48	(8.07, 42.34)

The maximum value of the log-likelihood function for the minimal model (with only two parameters, β_{02} and β_{03}) is -329.27 and for the fitted model (8.10) is -290.35, giving the likelihood ratio chi-squared statistic $C = 2\times(-290.35+329.27) = 77.84$ and pseudo $R^2 = (-329.27+290.35)/(-329.27) = 0.118$. The first statistic, which has 6 degrees of freedom (8 parameters in the fitted model minus 2 for the minimal model), is very significant compared with the $\chi^2(6)$ distribution, showing the overall importance of the explanatory variables. However, the second statistic suggests that only 11.8% of the 'variation' is 'explained' by these factors. From the Wald statistics $[b/\text{s.e.}(b)]$ and the odds ratios and the confidence intervals, it is clear that the importance of air-conditioning and power steering increased significantly with age. Also men considered these features less important than women did, although the statistical significance of this finding is dubious (especially considering the small frequencies in some cells).

To estimate the probabilities, first consider the preferences of women ($x_1 = 0$) aged 18-23 (so $x_2 = 0$ and $x_3 = 0$). For this group

$$\log\left(\frac{\widehat{\pi}_2}{\widehat{\pi}_1}\right) = -0.591 \quad \text{so} \quad \frac{\widehat{\pi}_2}{\widehat{\pi}_1} = e^{-0.591} = 0.5539,$$

$$\log\left(\frac{\widehat{\pi}_3}{\widehat{\pi}_1}\right) = -1.039 \quad \text{so} \quad \frac{\widehat{\pi}_3}{\widehat{\pi}_1} = e^{-1.039} = 0.3538$$

Table 8.3 *Results from fitting the nominal logistic regression model (8.10) to the data in Table 8.1.*

Sex	Age	Importance Rating*	Obs. freq.	Estimated probability	Fitted value	Pearson residual
Women	18-23	1	26	0.524	23.59	0.496
		2	12	0.290	13.07	-0.295
		3	7	0.186	8.35	-0.466
	24-40	1	9	0.234	10.56	-0.479
		2	21	0.402	18.07	0.690
		3	15	0.364	16.37	-0.340
	> 40	1	5	0.098	5.85	-0.353
		2	14	0.264	15.87	-0.468
		3	41	0.638	38.28	0.440
Men	18-23	1	40	0.652	42.41	-0.370
		2	17	0.245	15.93	0.267
		3	8	0.102	6.65	0.522
	24-40	1	17	0.351	15.44	0.396
		2	15	0.408	17.93	-0.692
		3	12	0.241	10.63	0.422
	> 40	1	8	0.174	7.15	0.320
		2	15	0.320	13.13	0.515
		3	18	0.505	20.72	-0.600
Total			300		300	
Sum of squares						3.931

* 1 denotes 'no/little' importance, 2 denotes 'important', 3 denotes 'very important'.

but $\widehat{\pi}_1+\widehat{\pi}_2+\widehat{\pi}_3 = 1$ so $\widehat{\pi}_1(1+0.5539+0.3538) = 1$, therefore $\widehat{\pi}_1 = 1/1.9077 = 0.524$ and hence $\widehat{\pi}_2 = 0.290$ and $\widehat{\pi}_3 = 0.186$. Now consider men ($x_1 = 1$) aged over 40 (so $x_2 = 0$, but $x_3 = 1$) so that $\log(\widehat{\pi}_2/\widehat{\pi}_1) = -0.591-0.388+1.588 = 0.609$, $\log(\widehat{\pi}_3/\widehat{\pi}_1) = 1.065$ and hence $\widehat{\pi}_1 = 0.174$, $\widehat{\pi}_2 = 0.320$ and $\widehat{\pi}_3 = 0.505$ (correct to 3 decimal places). These estimated probabilities can be multiplied by the total frequency for each sex $\times$ age group to obtain the 'expected' frequencies or fitted values. These are shown in Table 8.3, together with the Pearson residuals defined in (8.5). The sum of squares of the Pearson residuals, the chi-squared goodness of fit statistic (8.6), is $X^2 = 3.93$.

The maximal model that can be fitted to these data involves terms for age, sex and age $\times$ sex interactions. It has 6 parameters (a constant and coefficients for sex, two age categories and two age $\times$ sex interactions) for $j = 2$ and 6 parameters for $j = 3$, giving a total of 12 parameters. The maximum value of the log-likelihood function for the maximal model is -288.38. Therefore the deviance for the fitted model (8.10) is $D = 2\times(-288.38 + 290.35) = 3.94$. The degrees of freedom associated with this deviance are $12 - 8 = 4$ because the maximal model has 12 parameters and the fitted model has 8 parameters. As expected, the values of the goodness of fit statistics $D = 3.94$ and $X^2 = 3.93$ are very similar; when compared to the distribution $\chi^2(4)$ they suggest that model (8.10) provides a good description of the data.

An alternative model can be fitted with age group as covariate, that is

$$\log\left(\frac{\pi_j}{\pi_1}\right) = \beta_{0j} + \beta_{1j}x_1 + \beta_{2j}x_2; \quad j = 2, 3, \tag{8.11}$$

where

$$x_1 = \begin{cases} 1 & \text{for men} \\ 0 & \text{for women} \end{cases} \quad \text{and} \quad x_2 = \begin{cases} 0 & \text{for age group 18-23} \\ 1 & \text{for age group 24-40} \\ 2 & \text{for age group} > 40 \end{cases}$$

This model fits the data almost as well as (8.10) but with two fewer parameters. The maximum value of the log likelihood function is -291.05 so the difference in deviance from model (8.10) is

$$\triangle D = 2 \times (-290.35 + 291.05) = 1.4$$

which is not significant compared with the distribution $\chi^2(2)$. So on the grounds of parsimony model (8.11) is preferable.

8.4 Ordinal logistic regression

If there is an obvious natural order among the response categories then this can be taken into account in the model specification. The example on car preferences (Section 8.3.1) provides an illustration as the study participants rated the importance of air conditioning and power steering in four categories from 'not important' to 'very important'. Ordinal responses like this are common in areas such as market research, opinion polls and fields like psychiatry where 'soft' measures are common (Ashby et al., 1989).

In some situations there may, conceptually, be a continuous variable z which is difficult to measure, such as severity of disease. It is assessed by some crude method that amounts to identifying 'cut points', C_j, for the **latent variable** so that, for example, patients with small values are classified as having 'no disease', those with larger values of z are classified as having 'mild disease' or 'moderate disease' and those with high values are classified as having 'severe disease' (see Figure 8.2). The cutpoints $C_1, ..., C_{J-1}$ define J ordinal categories with associated probabilities $\pi_1, ..., \pi_J$ (with $\sum_{j=1}^{J} \pi_j = 1$).

Not all ordinal variables can be thought of in this way, because the underlying process may have many components, as in the car preference example. Nevertheless, the idea is helpful for interpreting the results from statistical models. For ordinal categories, there are several different commonly used models which are described in the next sections.

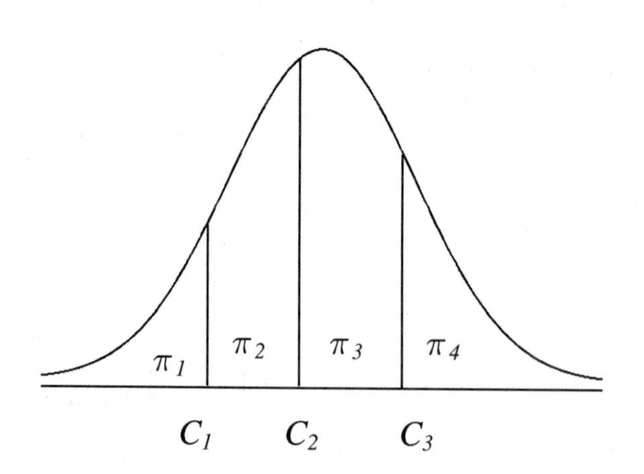

Figure 8.2 *Distribution of continuous latent variable and cutpoints that define an ordinal response variable.*

8.4.1 Cumulative logit model

The cumulative odds for the jth category is

$$\frac{P(z \leq C_j)}{P(z > C_j)} = \frac{\pi_1 + \pi_2 + ... + \pi_j}{\pi_{j+1} + ... + \pi_J};$$

see Figure 8.2. The cumulative logit model is

$$\log \frac{\pi_1 + ... + \pi_j}{\pi_{j+1} + ... + \pi_J} = \mathbf{x}_j^T \boldsymbol{\beta}_j. \tag{8.12}$$

8.4.2 Proportional odds model

If the linear predictor $\mathbf{x}_j^T \boldsymbol{\beta}_j$ in (8.12) has an intercept term β_{0j} which depends on the category j, but the other explanatory variables do not depend on j, then the model is

$$\log \frac{\pi_1 + ... + \pi_j}{\pi_{j+1} + ... + \pi_J} = \beta_{0j} + \beta_1 x_1 + ... + \beta_{p-1} x_{p-1}. \tag{8.13}$$

This is called the **proportional odds model**. It is based on the assumption that the effects of the covariates $x_1, ..., x_{p-1}$ are the same for all categories, on the logarithmic scale. Figure 8.3 shows the model for $J = 3$ response categories and one continuous explanatory variable x; on the log odds scale the probabilities for categories are represented by parallel lines.

As for the nominal logistic regression model (8.4), the odds ratio associated

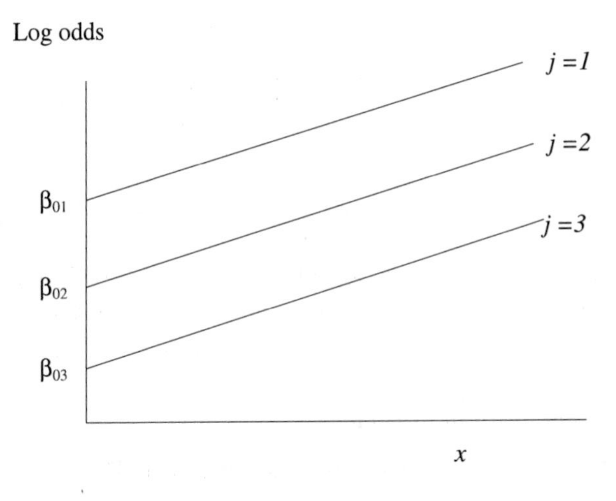

Figure 8.3 *Proportional odds model, on log odds scale.*

with an increase of one unit in an explanatory variable x_k is $\exp(\beta_k)$ where $k = 1, ..., p-1$.

If some of the categories are amalgamated, this does not change the parameter estimates $\beta_1, ..., \beta_{p-1}$ in (8.13) – although, of course, the terms β_{0j} will be affected (this is called the **collapsibility** property; see Ananth and Kleinbaum, 1997). This form of independence between the cutpoints C_j (in Figure 8.2) and the explanatory variables x_k is desirable for many applications.

Another useful property of the proportional odds model is that it is not affected if the labelling of the categories is reversed – only the signs of the parameters will be changed.

The appropriateness of the proportional odds assumption can be tested by comparing models (8.12) and (8.13), if there is only one explanatory variable x. If there are several explanatory variables the assumption can be tested separately for each variable by fitting (8.12) with the relevant parameter not depending on j.

The proportional odds model is the usual (or default) form of ordinal logistic regression provided by statistical software.

8.4.3 Adjacent categories logit model

One alternative to the cumulative odds model is to consider ratios of probabilities for successive categories, for example

$$\frac{\pi_1}{\pi_2}, \frac{\pi_2}{\pi_3}, ..., \frac{\pi_{J-1}}{\pi_J}.$$

The adjacent category logit model is

$$\log\left(\frac{\pi_j}{\pi_{j+1}}\right) = \mathbf{x}_j^T \boldsymbol{\beta}_j. \tag{8.14}$$

If this is simplified to

$$\log\left(\frac{\pi_j}{\pi_{j+1}}\right) = \beta_{0j} + \beta_1 x_1 + ... + \beta_{p-1} x_{p-1}$$

the effect of each explanatory variable is assumed to be the same for all adjacent pairs of categories. The parameters β_k are usually interpreted as odd ratios using $OR = \exp(\beta_k)$.

8.4.4 Continuation ratio logit model

Another alternative is to model the ratios of probabilities

$$\frac{\pi_1}{\pi_2}, \frac{\pi_1 + \pi_2}{\pi_3}, ..., \frac{\pi_1 + ... + \pi_{J-1}}{\pi_J}$$

or

$$\frac{\pi_1}{\pi_2 + ... + \pi_J}, \frac{\pi_2}{\pi_3 + ... + \pi_J}, ..., \frac{\pi_{J-1}}{\pi_J}.$$

The equation

$$\log\left(\frac{\pi_j}{\pi_{j+1} + ... + \pi_J}\right) = \mathbf{x}_j^T \boldsymbol{\beta}_j \tag{8.15}$$

models the odds of the response being in category j, i.e., $C_{j-1} < z \leq C_j$ conditional upon $z \geq C_{j-1}$. For example, for the car preferences data (Section 8.3.1), one could estimate the odds of respondents regarding air conditioning and power steering as 'unimportant' vs. 'important' and the odds of these features being 'very important' given that they are 'important' or 'very important', using

$$\log\left(\frac{\pi_1}{\pi_2 + \pi_3}\right) \quad \text{and} \quad \log\left(\frac{\pi_2}{\pi_3}\right).$$

This model may be easier to interpret than the proportional odds model if the probabilities for individual categories π_j are of interest (Agresti, 1996, Section 8.3.4).

8.4.5 Comments

Hypothesis tests for ordinal logistic regression models can be performed by comparing the fit of nested models or by using Wald statistics (or, less commonly, score statistics) based on the parameter estimates. Residuals and goodness of fit statistics are analogous to those for nominal logistic regression (Section 8.3).

The choice of model for ordinal data depends mainly on the practical problem being investigated. Comparisons of the models described in this chapter and some other models have been published by Holtbrugger and Schumacher (1991) and Ananth and Kleinbaum (1997), for example.

8.4.6 Example: Car preferences

The response variable for the car preference data is, of course, ordinal (Table 8.1). The following proportional odds model was fitted to these data:

$$\begin{aligned}\log\left(\frac{\pi_1}{\pi_2+\pi_3}\right) &= \beta_{01}+\beta_1x_1+\beta_2x_2+\beta_3x_3 \\ \log\left(\frac{\pi_1+\pi_2}{\pi_3}\right) &= \beta_{02}+\beta_1x_1+\beta_2x_2+\beta_3x_3 \qquad (8.16)\end{aligned}$$

where x_1, x_2 and x_3 are as defined for model (8.10).

The results are shown in Table 8.4. For model (8.16), the maximum value of the log-likelihood function is $l(\mathbf{b}) = -290.648$. For the minimal model, with only β_{01} and β_{02}, the maximum value is $l(\mathbf{b}_{\min}) = -329.272$ so, from (8.8), $C = 2\times(-290.648+329.272) = 77.248$ and, from (8.9), pseudo $R^2 = (-329.272+290.648)/(-329.272) = 0.117$.

The parameter estimates for the proportional odds model are all quite similar to those from the nominal logistic regression model (see Table 8.2). The estimated probabilities are also similar; for example, for females aged 18-23, $x_1 = 0, x_2 = 0$ and $x_3 = 0$ so, from (8.16), $\log\left(\frac{\pi_3}{\pi_1+\pi_2}\right) = -1.6550$ and $\log\left(\frac{\pi_2+\pi_3}{\pi_1}\right) = -0.0435$. If these equations are solved with $\pi_1+\pi_2+\pi_3 = 1$, the estimates are $\widehat{\pi}_1 = 0.5109, \widehat{\pi}_2 = 0.3287$ and $\widehat{\pi}_3 = 0.1604$. The probabilities for other covariate patterns can be estimated similarly and hence expected frequencies can be calculated, together with residuals and goodness of fit statistics. For the proportional odds model, $X^2 = 4.564$ which is consistent with distribution $\chi^2(7)$, indicating that the model described the data well (in this case $N = 18$, the maximal model has 12 parameters and model (8.13) has 5 parameters so degrees of freedom = 7).

For this example, the proportional odds logistic model for ordinal data and the nominal logistic model produce similar results. On the grounds of parsimony, model (8.16) would be preferred because it is simpler and takes into account the order of the response categories.

8.5 General comments

Although the models described in this chapter are developed from the logistic regression model for binary data, other link functions such as the probit or complementary log-log functions can also be used. If the response categories are regarded as crude measures of some underlying latent variable, z (as in Figure 8.2), then the optimal choice of the link function can depend on the shape of the distribution of z (McCullagh, 1980). Logits and probits are ap-

Table 8.4 *Results of proportional odds ordinal regression model (8.16) for the data in Table 8.1.*

Parameter	Estimate b	Standard error, s.e.(b)	Odds ratio OR (95% confidence interval)
β_{01}	-1.655	0.256	
β_{02}	-0.044	0.232	
β_1 : men	-0.576	0.226	0.56 (0.36, 0.88)
$\beta_2 : 24-40$	1.147	0.278	3.15 (1.83, 5.42)
$\beta_3 :> 40$	2.232	0.291	9.32 (5.28, 16.47)

propriate if the distribution is symmetric but the complementary log-log link may be better if the distribution is very skewed.

If there is doubt about the order of the categories then nominal logistic regression will usually be a more appropriate model than any of the models based on assumptions that the response categories are ordinal. Although the resulting model will have more parameters and hence fewer degrees of freedom and less statistical power, it may give results very similar to the ordinal models (as in the car preference example).

The estimation methods and sampling distributions used for inference depend on asymptotic results. For small studies, or numerous covariate patterns, each with few observations, the asymptotic results may be poor approximations.

Multicategory logistic models have only been readily available in statistical software from the 1990s. Their use has grown because the results are relatively easy to interpret provided that one variable can clearly be regarded as a response and the others as explanatory variables. If this distinction is unclear, for example, if data from a cross-sectional study are cross-tabulated, then log-linear models may be more appropriate. These are discussed in Chapter 9.

8.6 Exercises

8.1 If there are only $J = 2$ response categories, show that models (8.4), (8.12), (8.14) and (8.15) all reduce to the logistic regression model for binary data.

8.2 The data in Table 8.5 are from an investigation into satisfaction with housing conditions in Copenhagen (derived from Example W in Cox and Snell, 1981, from original data from Madsen, 1971). Residents in selected areas living in rented homes built between 1960 and 1968 were questioned about their satisfaction and the degree of contact with other residents. The data were tabulated by type of housing.

(a) Summarize the data using appropriate tables of percentages to show the associations between levels of satisfaction and contact with other residents, levels of satisfaction and type of housing, and contact and type of housing.

Table 8.5 *Satisfaction with housing conditions.*

	Satisfaction					
	Low		Medium		High	
Contact with other residents	Low	High	Low	High	Low	High
Tower block	65	34	54	47	100	100
Apartment	130	141	76	116	111	191
House	67	130	48	105	62	104

(b) Use nominal logistic regression to model associations between level of satisfaction and the other two variables. Obtain a parsimonious model that summarizes the patterns in the data.

(c) Do you think an ordinal model would be appropriate for associations between the levels of satisfaction and the other variables? Justify your answer. If you consider such a model to be appropriate, fit a suitable one and compare the results with those from (b).

(d) From the best model you obtained in (c), calculate the standardized residuals and use them to find where the largest discrepancies are between the observed frequencies and expected frequencies estimated from the model.

8.3 The data in Table 8.6 show tumor responses of male and female patients receiving treatment for small-cell lung cancer. There were two treatment regimes. For the sequential treatment, the same combination of chemotherapeutic agents was administered at each treatment cycle. For the alternating treatment, different combinations were alternated from cycle to cycle (data from Holtbrugger and Schumacher, 1991).

Table 8.6 *Tumor responses to two different treatments: numbers of patients in each category.*

Treatment	Sex	Progressive disease	No change	Partial remission	Complete remission
Sequential	Male	28	45	29	26
	Female	4	12	5	2
Alternating	Male	41	44	20	20
	Female	12	7	3	1

(a) Fit a proportional odds model to estimate the probabilities for each response category taking treatment and sex effects into account.

(b) Examine the adequacy of the model fitted in (a) using residuals and goodness of fit statistics.

(c) Use a Wald statistic to test the hypothesis that there is no difference in responses for the two treatment regimes.

(d) Fit two proportional odds models to test the hypothesis of no treatment difference. Compare the results with those for (c) above.

(e) Fit adjacent category models and continuation ratio models using logit, probit and complementary log-log link functions. How do the different models affect the interpretation of the results?

8.4 Consider ordinal response categories which can be interpreted in terms of continuous latent variable as shown in Figure 8.2. Suppose the distribution of this underlying variable is Normal. Show that the probit is the natural link function in this situation (Hint: see Section 7.3).

9

Count Data, Poisson Regression and Log-Linear Models

9.1 Introduction

The number of times an event occurs is a common form of data. Examples of **count** or **frequency** data include the number of tropical cyclones crossing the North Queensland coast (Section 1.6.5) or the numbers of people in each cell of a contingency table summarizing survey responses (e.g., satisfaction ratings for housing conditions, Exercise 8.2).

The **Poisson distribution** is often used to model count data. If Y is the number of occurrences, its probability distribution can be written as

$$f(y) = \frac{\mu^y e^{-\mu}}{y!}, \quad y = 0, 1, 2, ...$$

where μ is the average number of occurrences. It can be shown that $E(Y) = \mu$ and $\text{var}(Y) = \mu$ (see Exercise 3.4).

The parameter μ requires careful definition. Often it needs to be described as a rate; for example, the average number of customers who buy a particular product out of every 100 customers who enter the store. For motor vehicle crashes the rate parameter may be defined in many different ways: crashes per 1,000 population, crashes per 1,000 licensed drivers, crashes per 1,000 motor vehicles, or crashes per 100,000 kms travelled by motor vehicles. The time scale should be included in the definition; for example, the motor vehicle crash rate is usually specified as the rate per year (e.g., crashes per 100,000 kms per year), while the rate of tropical cyclones refers to the cyclone season from November to April in Northeastern Australia. More generally, the rate is specified in terms of units of 'exposure'; for instance, customers entering a store are 'exposed' to the opportunity to buy the product of interest. For occupational injuries, each worker is exposed for the period he or she is at work, so the rate may be defined in terms of person-years 'at risk'.

The effect of explanatory variables on the response Y is modelled through the parameter μ. This chapter describes models for two situations.

In the first situation, the events relate to varying amounts of 'exposure' which need to be taken into account when modelling the rate of events. **Poisson regression** is used in this case. The other explanatory variables (in addition to 'exposure') may be continuous or categorical.

In the second situation, 'exposure' is constant (and therefore not relevant to the model) and the explanatory variables are usually categorical. If there are only a few explanatory variables the data are summarized in a cross-classified table. The response variable is the frequency or count in each cell of the table. The variables used to define the table are all treated as explanatory variables. The study design may mean that there are some constraints on the cell frequencies (for example, the totals for each row of the table may be equal) and these need to be taken into account in the modelling. The term **log-linear model**, which basically describes the role of the link function, is used for the generalized linear models appropriate for this situation.

The next section describes Poisson regression. A numerical example is used to illustrate the concepts and methods, including model checking and inference. Subsequent sections describe relationships between probability distributions for count data, constrained in various ways, and the log-linear models that can be used to analyze the data.

9.2 Poisson regression

Let $Y_1, ..., Y_N$ be independent random variables with Y_i denoting the number of events observed from exposure n_i for the ith covariate pattern. The expected value of Y_i can be written as

$$E(Y_i) = \mu_i = n_i\theta_i.$$

For example, suppose Y_i is the number of insurance claims for a particular make and model of car. This will depend on the number of cars of this type that are insured, n_i, and other variables that affect θ_i, such as the age of the cars and the location where they are used. The subscript i is used to denote the different combinations of make and model, age, location and so on.

The dependence of θ_i on the explanatory variables is usually modelled by

$$\theta_i = e^{\mathbf{x}_i^T\boldsymbol{\beta}} \tag{9.1}$$

Therefore the generalized linear model is

$$E(Y_i) = \mu_i = n_i e^{\mathbf{x}_i^T\boldsymbol{\beta}}; \quad Y_i \sim \text{Poisson } (\mu_i). \tag{9.2}$$

The natural link function is the logarithmic function

$$\log\mu_i = \log n_i + \mathbf{x}_i^T\boldsymbol{\beta}. \tag{9.3}$$

Equation (9.3) differs from the usual specification of the linear component due to the inclusion of the term $\log n_i$. This term is called the **offset**. It is a known constant which is readily incorporated into the estimation procedure. As usual, the terms $\mathbf{x}_i$ and $\boldsymbol{\beta}$ describe the covariate pattern and parameters, respectively.

For a binary explanatory variable denoted by an indictor variable, $x_j = 0$ if the factor is absent and $x_j = 1$ if it is present, the **rate ratio**, RR, for presence vs. absence is

$$RR = \frac{\mathrm{E}(Y_i \mid present)}{\mathrm{E}(Y_i \mid absent)} = e^{\beta_j}$$

from (9.1), provided all the other explanatory variables remain the same. Similarly, for a continuous explanatory variable x_k, a one-unit increase will result in a multiplicative effect of e^{β_k} on the rate μ. Therefore, parameter estimates are often interpreted on the exponential scale e^{β} in terms of ratios of rates.

Hypotheses about the parameters β_j can be tested using the Wald, score or likelihood ratio statistics. Confidence intervals can be estimated similarly. For example, for parameter β_j

$$\frac{b_j - \beta_j}{s.e.(b_j)} \sim N(0,1) \tag{9.4}$$

approximately. Alternatively, hypothesis testing can be performed by comparing the goodness of fit of appropriately defined nested models (see Chapter 4).

The fitted values are given by

$$\widehat{Y}_i = \widehat{\mu}_i = n_i e^{\mathbf{x}_i^T \mathbf{b}}, \quad i = 1, ..., N.$$

These are often denoted by e_i because they are estimates of the expected values $\mathrm{E}(Y_i) = \mu_i$. As $\mathrm{var}(Y_i) = \mathrm{E}(Y_i)$ for the Poisson distribution, the standard error of Y_i is estimated by $\sqrt{e_i}$ so the **Pearson residuals** are

$$r_i = \frac{o_i - e_i}{\sqrt{e_i}} \tag{9.5}$$

where o_i denotes the observed value of Y_i. As outlined in Section 6.2.6, these residuals may be further refined to

$$r_{pi} = \frac{o_i - e_i}{\sqrt{e_i}\sqrt{1-h_i}}$$

where the leverage, h_i, is the ith element on the diagonal of the hat matrix.

For the Poisson distribution, the residuals given by (9.5) and the chi-squared goodness of fit statistic are related by

$$X^2 = \sum r_i^2 = \sum \frac{(o_i - e_i)^2}{e_i}$$

which is the usual definition of the chi-squared statistic for contingency tables.

The deviance for a Poisson model is given in Section 5.6.3. It can be written in the form

$$D = 2\sum \left[o_i \log(o_i/e_i) - (o_i - e_i)\right]. \tag{9.6}$$

However for most models $\sum o_i = \sum e_i$, see Exercise 9.1, so the deviance simplifies to

$$D = 2\sum \left[o_i \log(o_i/e_i)\right]. \tag{9.7}$$

The **deviance residuals** are the components of D in (9.6),

$$d_i = \mathrm{sign}(o_i - e_i)\sqrt{2\left[o_i \log(o_i/e_i) - (o_i - e_i)\right]}, \quad i = 1, ..., N \tag{9.8}$$

so that $D = \sum d_i^2$.

The goodness of fit statistics X^2 and D are closely related. Using the Taylor series expansion given in Section 7.5,

$$o \log(\frac{o}{e}) = (o - e) + \frac{1}{2}\frac{(o-e)^2}{e} + ...$$

so that, approximately, from (9.6)

$$\begin{aligned} D &= 2\sum \left[(o_i - e_i) + \frac{1}{2}\frac{(o_i - e_i)^2}{e_i} - (o_i - e_i) \right] \\ &= \sum \frac{(o_i - e_i)^2}{e_i} = X^2. \end{aligned}$$

The statistics D and X^2 can be used directly as measures of goodness of fit, as they can be calculated from the data and the fitted model (because they do not involve any nuisance parameters like σ^2 for the Normal distribution). They can be compared with the central chi-squared distribution with $N - p$ degrees of freedom, where p is the number of parameters that are estimated. The chi-squared distribution is likely to be a better approximation for the sampling distribution of X^2 than for the sampling distribution of D (see Section 7.5).

Two other summary statistics provided by some software are the likelihood ratio chi-squared statistic and pseudo-R^2. These are based on comparisons between the maximum value of the log-likelihood function for a minimal model with no covariates, $\log \mu_i = \log n_i + \beta_1$, and the maximum value of the log-likelihood function for model (9.3) with p parameters. The likelihood ratio chi-squared statistic $C = 2\left[l(\mathbf{b}) - l(\mathbf{b}_{\min})\right]$ provides an overall test of the hypotheses that $\beta_2 = ... = \beta_p = 0$, by comparison with the central chi-squared distribution with $p - 1$ degrees of freedom (see Exercise 7.4). Less formally, pseudo $R^2 = \left[l(\mathbf{b}_{\min}) - l(\mathbf{b})\right]/l(\mathbf{b}_{\min})$ provides an intuitive measure of fit.

Other diagnostics, such as delta-betas and related statistics, are also available for Poisson models.

9.2.1 Example of Poisson regression: British doctors' smoking and coronary death

The data in Table 9.1 are from a famous study conducted by Sir Richard Doll and colleagues. In 1951, all British doctors were sent a brief questionnaire about whether they smoked tobacco. Since then information about their deaths has been collected. Table 9.1 shows the numbers of deaths from coronary heart disease among male doctors 10 years after the survey. It also shows the total number of person-years of observation at the time of the analysis (Breslow and Day, 1987: Appendix 1A and page 112).

The questions of interest are:

1. Is the death rate higher for smokers than non-smokers?
2. If so, by how much?
3. Is the differential effect related to age?

Table 9.1 *Deaths from coronary heart disease after 10 years among British male doctors categorized by age and smoking status in 1951.*

Age group	Smokers		Non-smokers	
	Deaths	Person-years	Deaths	Person-years
35 – 44	32	52407	2	18790
45 – 54	104	43248	12	10673
55 – 64	206	28612	28	5710
65 – 74	186	12663	28	2585
75 – 84	102	5317	31	1462

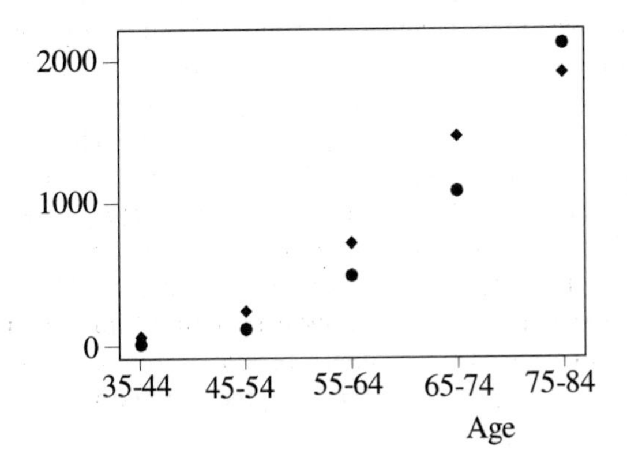

Figure 9.1 *Death rates from coronary heart disease per 100,000 person-years for smokers (diamonds) and non-smokers (dots).*

Figure 9.1 shows the death rates per 100,000 person-years from coronary heart disease for smokers and non-smokers. It is clear that the rates increase with age but more steeply than in a straight line. Death rates appear to be generally higher among smokers than non-smokers but they do not rise as rapidly with age. Various models can be specified to describe these data well (see Exercise 9.2). One model, in the form of (9.3) is

$$\log(deaths_i) = \log(population_i) + \beta_1 + \beta_2 smoke_i + \beta_3 agecat_i + \beta_4 agesq_i + \beta_5 smkage_i \quad (9.9)$$

where the subscript i denotes the ith subgroup defined by age group and smoking status ($i = 1, ..., 5$ for ages $35 - 44, ..., 75 - 84$ for smokers and $i = 6, ..., 10$ for the corresponding age groups for non-smokers). The term $deaths_i$ denotes the expected number of deaths and $population_i$ denotes the number of doctors at risk in group i. For the other terms, $smoke_i$ is equal to one for smokers and zero for non-smokers; $agecat_i$ takes the values $1, ..., 5$ for age groups $35 - 44, ..., 75 - 84$; $agesq_i$ is the square of $agecat_i$ to take account of the non-linearly of the rate of increase; and $smkage_i$ is equal to $agecat_i$ for smokers and zero for non-smokers, thus describing a differential rate of increase with age.

Table 9.2 shows the parameter estimates in the form of rate ratios $e^{\widehat{\beta}_j}$. The Wald statistics (9.4) to test $\beta_j = 0$ all have very small p-values and the 95% confidence intervals for e^{β_j} do not contain unity showing that all the terms are needed in the model. The estimates show that the risk of coronary deaths was, on average, about 4 times higher for smokers than non-smokers (based on the rate ratio for *smoke*), after the effect of age is taken into account. However, the effect is attenuated as age increases (coefficient for *smkage*). Table 9.3 shows that the model fits the data very well; the expected number of deaths estimated from (9.9) are quite similar to the observed numbers of deaths and so the Pearson residuals calculated from (9.5) and deviance residuals from (9.8) are very small.

For the minimal model, with only the parameter β_1, the maximum value for the log-likelihood function is $l(b_{\min}) = -495.067$. The corresponding value for model (9.9) is $l(\mathbf{b}) = -28.352$. Therefore, an overall test of the model (testing $\beta_j = 0$ for $j = 2, ..., 5$) is $C = 2\left[l(\mathbf{b}) - l(b_{\min})\right] = 933.43$ which is highly statistically significant compared to the chi-squared distribution with 4 degrees of freedom. The pseudo R^2 value is 0.94, or 94%, which suggests a good fit. More formal tests of the goodness of fit are provided by the statistics $X^2 = 1.550$ and $D = 1.635$ which are small compared to the chi-squared distribution with $N - p = 10 - 5 = 5$ degree of freedom.

Table 9.2 *Parameter estimates obtained by fitting model (9.9) to the data in Table 9.1.*

Term	*agecat*	*agesq*	*smoke*	*smkage*
$\widehat{\beta}$	2.376	-0.198	1.441	-0.308
$s.e.(\widehat{\beta})$	0.208	0.027	0.372	0.097
Wald statistic	11.43	-7.22	3.87	-3.17
p-value	<0.001	<0.001	<0.001	0.002
Rate ratio	10.77	0.82	4.22	0.74
95% confidence interval	7.2, 16.2	0.78, 0.87	2.04, 8.76	0.61, 0.89

9.3 Examples of contingency tables

Before specifying log-linear models for frequency data summarized in contingency tables, it is important to consider how the design of the study may determine constraints on the data. The study design also affects the choice

Table 9.3 *Observed and estimated expected numbers of deaths and residuals for the model described in Table 9.2.*

Age category	Smoking category	Observed deaths	Expected deaths	Pearson residual	Deviance residual
1	1	32	29.58	0.444	0.438
2	1	104	106.81	-0.272	-0.273
3	1	206	208.20	-0.152	-0.153
4	1	186	182.83	0.235	0.234
5	1	102	102.58	-0.057	-0.057
1	0	2	3.41	-0.766	-0.830
2	0	12	11.54	0.135	0.134
3	0	28	27.74	0.655	0.641
4	0	28	30.23	-0.405	-0.411
5	0	31	31.07	-0.013	-0.013
sum of squares*				1.550	1.635

* calculated from residuals correct to more significant figures than shown here.

of probability models to describe the data. These issues are illustrated in the following three examples.

9.3.1 *Example: Cross-sectional study of malignant melanoma*

These data are from a cross-sectional study of patients with a form of skin cancer called malignant melanoma. For a sample of $n = 400$ patients, the site of the tumor and its histological type were recorded. The data, numbers of patients with each combination of tumor type and site, are given in Table 9.4.

Table 9.4 *Malignant melanoma: frequencies for tumor type and site (Roberts et al., 1981).*

	Site			
Tumor type	Head & neck	Trunk	Extrem -ities	Total
Hutchinson's melanotic freckle	22	2	10	34
Superficial spreading melanoma	16	54	115	185
Nodular	19	33	73	125
Indeterminate	11	17	28	56
Total	68	106	226	400

The question of interest is whether there is any association between tumor type and site. Table 9.5 shows the data displayed as percentages of row and column totals. It appears that Hutchinson's melanotic freckle is more common on the head and neck but there is little evidence of association between other tumor types and sites.

Table 9.5 *Malignant melanoma: row and column percentages for tumor type and site.*

	Site			
Tumor type	Head & neck	Trunk	Extrem -ities	Total
Row percentages				
Hutchinson's melanotic freckle	64.7	5.9	29.4	100
Superficial spreading melanoma	8.6	29.2	62.2	100
Nodular	15.2	26.4	58.4	100
Indeterminate	19.6	30.4	50.0	100
All types	17.0	26.5	56.5	100
Column percentages				
Hutchinson's melanotic freckle	32.4	1.9	4.4	8.50
Superficial spreading melanoma	23.5	50.9	50.9	46.25
Nodular	27.9	31.1	32.3	31.25
Indeterminate	16.2	16.0	12.4	14.00
All types	100.0	99.9	100.0	100.0

Let Y_{jk} denote the frequency for the (j,k)th cell with $j = 1, ..., J$ and $k = 1, ..., K$. In this example, there are $J = 4$ rows, $K = 3$ columns and the constraint that $\sum_{j=1}^{J} \sum_{k=1}^{K} Y_{jk} = n$, where $n = 400$ is fixed by the design of the study. If the Y_{jk}'s are independent random variables with Poisson distributions with parameters $\mathrm{E}(Y_{jk}) = \mu_{jk}$, then their sum has the Poisson distribution with parameter $\mathrm{E}(n) = \mu = \sum\sum \mu_{jk}$. Hence the joint probability distribution of the Y_{jk}'s, conditional on their sum n, is the multinomial distribution

$$f(\mathbf{y}\,|n) = n! \prod_{j=1}^{J} \prod_{k=1}^{K} \theta_{jk}^{y_{jk}} / y_{jk}!$$

where $\theta_{jk} = \mu_{jk}/\mu$. This result is derived in Section 8.2. The sum of the terms θ_{jk} is unity because $\sum\sum \mu_{jk} = \mu$; also $0 < \theta_k < 1$. Thus θ_{jk} can be interpreted as the probability of an observation in the (j,k)th cell of the table. Also the expected value of Y_{jk} is

$$\mathrm{E}(Y_{jk}) = \mu_{jk} = n\theta_{jk}.$$

The usual link function for a Poisson model gives

$$\log \mu_{jk} = \log n + \log \theta_{jk}$$

which is like equation (9.3), except that the term $\log n$ is the same for all the Y_{jk}'s.

9.3.2 *Example: Randomized controlled trial of influenza vaccine*

In a prospective study of a new living attenuated recombinant vaccine for influenza, patients were randomly allocated to two groups, one of which was given the new vaccine and the other a saline placebo. The responses were titre levels of hemagglutinin inhibiting antibody found in the blood six weeks after vaccination; they were categorized as 'small', 'medium' or 'large'. The cell frequencies in the rows of Table 9.6 are constrained to add to the number of subjects in each treatment group (35 and 38 respectively). We want to know if the pattern of responses is the same for each treatment group.

Table 9.6 *Flu vaccine trial.*

	Response			
	Small	Moderate	Large	Total
Placebo	25	8	5	38
Vaccine	6	18	11	35

(Data from R.S. Gillett, personal communication)

In this example the row totals are fixed. Thus the joint probability distribution for each row is multinomial

$$f(y_{j1}, y_{j2}, ..., y_{jK} \,| y_{j.}) = y_{j.}! \prod_{k=1}^{K} \theta_{jk}^{y_{jk}} y_{jk}!,$$

where $y_{j.} = \sum_{k=1}^{K} y_{jk}$ is the row total and $\sum_{k=1}^{K} \theta_{jk} = 1$. So the joint probability distribution for all the cells in the table is the **product multinomial distribution**

$$f(\mathbf{y} \,| y_{1.}, y_{2.}, ..., y_{J.}) = \prod_{j=1}^{J} y_{j.}! \prod_{k=1}^{K} \theta_{jk}^{y_{jk}} y_{jk}!$$

where $\sum_{k=1}^{K} \theta_{jk} = 1$ for each row. In this case $E(Y_{jk}) = y_{j.}\theta_{jk}$ so that

$$\log E(Y_{jk}) = \log \mu_{jk} = \log y_{j.} + \log \theta_{jk}.$$

If the response pattern was the same for both groups then $\theta_{jk} = \theta_{.k}$ for $k = 1, ..., K$.

9.3.3 *Example: Case control study of gastric and duodenal ulcers and aspirin use*

In this retrospective case-control study a group of ulcer patients was compared to a group of control patients not known to have peptic ulcer, but who were similar to the ulcer patients with respect to age, sex and socio-economic status. The ulcer patients were classified according to the site of the ulcer: gastric or duodenal. Aspirin use was ascertained for all subjects. The results are shown in Table 9.7.

Table 9.7 *Gastric and duodenal ulcers and aspirin use: frequencies (Duggan et al., 1986).*

	Aspirin use		
	Non-user	User	Total
Gastric ulcer			
Control	62	6	68
Cases	39	25	64
Duodenal ulcer			
Control	53	8	61
Cases	49	8	57

This is a $2 \times 2 \times 2$ contingency table. Some questions of interest are:

1. Is gastric ulcer associated with aspirin use?
2. Is duodenal ulcer associated with aspirin use?
3. Is any association with aspirin use the same for both ulcer sites?

   When the data are presented as percentages of row totals (Table 9.8) it appears that aspirin use is more common among ulcer patients than among controls for gastric ulcer but not for duodenal ulcer.

In this example, the numbers of patients with each type of ulcer and the numbers in each of the groups of controls; that is, the four row totals in Table 9.7 were all fixed.

Let $j = 1$ or 2 denote the controls or cases, respectively; $k = 1$ or 2 denote gastric ulcers or duodenal ulcers, respectively; and $l = 1$ for patients who did

Table 9.8 *Gastric and duodenal ulcers and aspirin use: row percentages for the data in Table 9.7.*

	Aspirin use		
	Non-user	User	Total
Gastric ulcer			
Control	91	9	100
Cases	61	39	100
Duodenal ulcer			
Control	87	13	100
Cases	86	14	100

not use aspirin and $l = 2$ for those who did. In general, let Y_{jkl} denote the frequency of observations in category (j, k, l) with $j = 1, ..., J, k = 1, ..., K$ and $l = 1, ..., L$. If the marginal totals $y_{jk.}$ are fixed, the joint probability distribution for the Y_{jkl}'s is

$$f(\mathbf{y}\,|y_{11.}, ..., y_{JK.}) = \prod_{j=1}^{J}\prod_{k=1}^{K} y_{jk.}! \prod_{l=1}^{L} \theta_{jkl}^{y_{jkl}} y_{jkl}!$$

where $\mathbf{y}$ is the vector of Y_{jkl}'s and $\sum_l \theta_{jkl} = 1$ for $j = 1, ..., J$ and $k = 1, ..., K$. This is another form of **product multinomial distribution**. In this case, $\mathrm{E}(Y_{jkl}) = \mu_{jkl} = y_{jk.}\theta_{jkl}$, so that

$$\log \mu_{jkl} = \log y_{jk.} + \log \theta_{jkl}.$$

9.4 Probability models for contingency tables

The examples in Section 9.3 illustrate the main probability models for contingency table data. In general, let the vector $\mathbf{y}$ denote the frequencies Y_i in N cells of a cross-classified table.

9.4.1 Poisson model

If there were no constraints on the Y_i's they could be modelled as independent random variables with the parameters $\mathrm{E}(Y_i) = \mu_i$ and joint probability distribution

$$f(\mathbf{y}; \boldsymbol{\mu}) = \prod_{i=1}^{N} \mu_i^{y_i} e^{-\mu_i} y_i!$$

where $\boldsymbol{\mu}$ is a vector of μ_i's.

9.4.2 Multinomial model

If the only constraint is that the sum of the Y_i's is n, then the following multinomial distribution may be used

$$f(\mathbf{y}; \boldsymbol{\mu}\,|n) = n! \prod_{i=1}^{N} \theta_i^{y_i} y_i!$$

where $\sum_{i=1}^{N} \theta_i = 1$ and $\sum_{i=1}^{N} y_i = n$. In this case, $\mathrm{E}(Y_i) = n\theta_i$.

For a two dimensional contingency table (like Table 9.4 for the melanoma data), if j and k denote the rows and columns then the most commonly considered hypothesis is that the row and column variables are independent so that

$$\theta_{jk} = \theta_{j.}\theta_{.k}$$

where $\theta_{j.}$ and $\theta_{.k}$ denote the marginal probabilities with $\sum_j \theta_{j.} = 1$ and

$\sum_k \theta_{.k} = 1$. This hypothesis can be tested by comparing the fit of two linear models for the logarithm of $\mu_{jk} = \mathrm{E}(Y_{jk})$; namely

$$\log \mu_{jk} = \log n + \log \theta_{jk}$$

and

$$\log \mu_{jk} = \log n + \log \theta_{j.} + \log \theta_{.k} \ .$$

9.4.3 Product multinomial models

If there are more fixed marginal totals than just the overall total n, then appropriate products of multinomial distributions can be used to model the data.

For example, for a three dimensional table with J rows, K columns and L layers, if the row totals are fixed in each layer the joint probability for the Y_{jkl}'s is

$$f(\mathbf{y}|y_{j.l}, j = 1, ..., J, l = 1, ..., L) = \prod_{j=1}^{J}\prod_{l=1}^{L} y_{j.l}! \prod_{k=1}^{K} \theta_{jkl}^{y_{jkl}} y_{jkl}!$$

where $\sum_k \theta_{jkl} = 1$ for each combination of j and l. In this case, $\mathrm{E}(Y_{jkl}) = y_{j.l}\theta_{jkl}$.

If only the layer totals are fixed, then

$$f(\mathbf{y}|y_{..l}, l = 1, ..., L) = \prod_{l=1}^{L} y_{..l}! \prod_{j=1}^{J}\prod_{k=1}^{K} \theta_{jkl}^{y_{jkl}} y_{jkl}!$$

with$\sum_j \sum_k \theta_{jkl} = 1$ for $l = 1, ..., L$. Also $\mathrm{E}(Y_{jkl}) = y_{..l}\theta_{jkl}$.

9.5 Log-linear models

All the probability models given in Section 9.4 are based on the Poisson distribution and in all cases $\mathrm{E}(Y_i)$ can be written as a product of parameters and other terms. Thus, the natural link function for the Poisson distribution, the logarithmic function, yields a linear component

$$\log \mathrm{E}(Y_i) = \text{constant} + \mathbf{x}_i^T \boldsymbol{\beta}.$$

The term **log-linear model** is used to describe all these generalized linear models.

For the melanoma Example 9.3.1, if there are no associations between site and type of tumor so that these two variables are independent, their joint probability θ_{jk} is the product of the marginal probabilities

$$\theta_{jk} = \theta_{j.}\theta_{.k} \quad , \quad j = 1, ..., J \text{ and } k = 1, ..., K.$$

The hypothesis of independence can be tested by comparing the additive model (on the logarithmic scale)

$$\log \mathrm{E}(Y_{jk}) = \log n + \log \theta_{j.} + \log \theta_{.k} \qquad (9.10)$$

with the model

$$\log E(Y_{jk}) = \log n + \log \theta_{jk}. \tag{9.11}$$

This is analogous to analysis of variance for a two factor experiment without replication (see Section 6.4.2). Equation (9.11) can be written as the saturated model

$$\log E(Y_{jk}) = \mu + \alpha_j + \beta_k + (\alpha\beta)_{jk}$$

and equation (9.10) can be written as the additive model

$$\log E(Y_{jk}) = \mu + \alpha_j + \beta_k.$$

Since the term $\log n$ has to be in all models, the minimal model is

$$\log E(Y_{jk}) = \mu.$$

For the flu vaccine trial, Example 9.3.2, $E(Y_{jk}) = y_{j.}\theta_{jk}$ if the distribution of responses described by the θ_{jk}'s differs for the j groups, or $E(Y_{jk}) = y_{j.}\theta_{.k}$ if it is the same for all groups. So the hypothesis of **homogeneity** of the response distributions can be tested by comparing the model

$$\log E(Y_{jk}) = \mu + \alpha_j + \beta_k + (\alpha\beta)_{jk},$$

corresponding to $E(Y_{jk}) = y_{j.}\theta_{jk}$, and the model

$$\log E(Y_{jk}) = \mu + \alpha_j + \beta_k$$

corresponding to $E(Y_{jk}) = y_{j.}\theta_{.k}$. The minimal model for these data is

$$\log E(Y_{jk}) = \mu + \alpha_j$$

because the row totals, corresponding to the subscript j, are fixed by the design of the study.

More generally, the specification of the linear components for log-linear models bears many resemblances to the specification for ANOVA models. The models are **hierarchical,** meaning that if a higher-order (interaction) term is included in the model then all the related lower-order terms are also included. Thus, if the two-way (first-order) interaction $(\alpha\beta)_{jk}$ is included then so are the main effects α_j and β_k and the constant μ. Similarly, if second-order interactions $(\alpha\beta\gamma)_{jkl}$ are included then so are the first-order interactions $(\alpha\beta)_{jk}, (\alpha\gamma)_{jl}$ and $(\beta\gamma)_{kl}$.

If log-linear models are specified analogously to ANOVA models, they include too many parameters so that sum-to-zero or corner-point constraints are needed. Interpretation of the parameters is usually simpler if reference or corner-point categories are identified so that parameter estimates describe effects for other categories relative to the reference categories.

For contingency tables the main questions almost always relate to associations between variables. Therefore, in log-linear models, the terms of primary interest are the interactions involving two or more variables.

9.6 Inference for log-linear models

Although three types of probability distributions are used to describe contingency table data (see Section 9.4), Birch (1963) showed that for any log-linear model the maximum likelihood estimators are the same for all these distributions provided that the parameters which correspond to the fixed marginal totals are always included in the model. This means that for the purpose of estimation, the Poisson distribution can always be assumed. As the Poisson distribution belongs to the exponential family and the parameter constraints can be incorporated into the linear component, all the standard methods for generalized linear models can be used.

The adequacy of a model can be assessed using the goodness of fit statistics X^2 or D (and sometimes C and pseudo R^2) summarized in Section 9.2 for Poisson regression. More insight into model adequacy can often be obtained by examining the Pearson or deviance residuals given by equations (9.5) and (9.8) respectively. Hypothesis tests can be conducted by comparing the difference in goodness of fit statistics between a general model corresponding to an alternative hypothesis and a nested, simpler model corresponding to a null hypothesis.

These methods are illustrated in the following examples.

9.7 Numerical examples

9.7.1 Cross-sectional study of malignant melanoma

For the data in Table 9.4 the question of interest is whether there is an association between tumor type and site. This can be examined by testing the null hypothesis that the variables are independent.

The conventional chi-squared test of independence for a two dimensional table is performed by calculating expected frequencies for each cell based on the marginal totals, $e_{jk} = y_{j.}y_{.k}/n$, calculating the chi-squared statistic $X^2 = \sum_j \sum_k (y_{jk} - e_{jk})^2 / e_{jk}$ and comparing this with the central chi-squared distribution with $(J-1)(K-1)$ degrees of freedom. The observed and expected frequencies are shown in Table 9.9. These give

$$X^2 = \frac{(22-5.78)^2}{5.78} + ... + \frac{(28-31.64)^2}{31.64} = 65.8.$$

The value $X^2 = 65.8$ is very significant compared to the $\chi^2(6)$ distribution. Examination of the observed frequencies y_{jk} and expected frequencies e_{jk} shows that Hutchinson's melanotic freckle is more common on the head and neck than would be expected if site and type were independent.

The corresponding analysis using log-linear models involves fitting the additive model (9.10) corresponding to the hypothesis of independence. The saturated model (9.11) and the minimal model with only a term for the mean effect are also fitted for illustrative purposes. The results for all three models are shown in Table 9.10. For the reference category of Hutchinson's melanotic

Table 9.9 *Conventional chi-squared test of independence for melanoma data in Table 9.4; expected frequencies are shown in brackets.*

	Site			
Tumor type	Head & Neck	Trunk	Extrem -ities	Total
Hutchinson's melanotic freckle	22 (5.78)	2 (9.01)	10 (19.21)	34
Superficial spreading melanoma	16 (31.45)	54 (49.03)	115 (104.52)	185
Nodular	19 (21.25)	33 (33.13)	73 (70.62)	125
Indeterminate	11 (9.52)	17 (14.84)	28 (31.64)	56
Total	68	106	226	400

freckle (HMF) on the head or neck (HNK), the expected frequencies are as follows:

minimal model: $e^{3.507} = 33.35$;

additive model: $e^{1.754} = 5.78$, as in Table 9.9;

saturated model: $e^{3.091} = 22$, equal to observed frequency.

For indeterminate tumors (IND) in the extremities (EXT), the expected frequencies are:

minimal model: $e^{3.507} = 33.35$;

additive model: $e^{1.754+0.499+1.201} = 31.64$, as in Table 9.9;

saturated model: $e^{3.091-0.693-0.788+1.723} = 28$, equal to observed frequency.

The saturated model with 12 parameters fits the 12 data points exactly. The additive model corresponds to the conventional analysis. The deviance for the additive model can be calculated from the sum of squares of the deviance residuals given by (9.8), or from twice the difference between the maximum values of the log-likelihood function for this model and the saturated model, $\triangle D = 2[-29.556 - (-55.453)] = 51.79$.

For this example, the conventional chi-squared test for independence and log-linear modelling produce exactly the same results. The advantage of log-linear modelling is that it provides a method for analyzing more complicated cross-tabulated data as illustrated by the next example.

9.7.2 *Case control study of gastric and duodenal ulcer and aspirin use*

Preliminary analysis of the 2 × 2 tables for gastric ulcer and duodenal ulcer separately suggests that aspirin use may be a risk factor for gastric ulcer but not for duodenal ulcer. For analysis of the full data set, Table 9.7, the main effects for case-control status (CC), ulcer site (GD) and the interaction

Table 9.10 *Log-linear models for the melanoma data in Table 9.4; coefficients, b, with standard errors in brackets.*

Term *	Saturated model (9.10)	Additive model (9.9)	Minimal model
Constant	3.091 (0.213)	1.754 (0.204)	3.507 (0.05)
SSM	-0.318 (0.329)	1.694 (0.187)	
NOD	-0.147 (0.313)	1.302 (0.193)	
IND	-0.693 (0.369)	0.499 (0.217)	
TNK	-2.398 (0.739)	0.444 (0.155)	
EXT	-0.788 (0.381)	1.201 (0.138)	
$SSM * TNK$	3.614 (0.792)		
$SSM * EXT$	2.761 (0.465)		
$NOD * TNK$	2.950 (0.793)		
$NOD * EXT$	2.134 (0.460)		
$IND * TNK$	2.833 (0.834)		
$IND * EXT$	1.723 (0.522)		
log-likelihood	-29.556	-55.453	-177.16
X^2	0.0	65.813	
D	0.0	51.795	

*Reference categories are: Hutchinson's melanotic freckle (HMF) and head and neck (HNK). Other categories are: for type, superficial spreading melanoma (SSM), nodular (NOD) and indeterminate (IND); for site, trunk (TNK) and extremities (EXT).

between these terms ($CC \times GD$) have to be included in all models (as these correspond to the fixed marginal totals). Table 9.11 shows the results of fitting this and several more complex models involving aspirin use (AP).

The comparison of aspirin use between cases and controls can be summarized by the deviance difference for the second and third rows of Table 9.11

$$\triangle D = 2[-25.08 - (-30.70)] = 11.24.$$

This value is statistically significant compared with the $\chi^2(1)$ distribution, suggesting that aspirin is a risk factor for ulcers. Comparison between the third and fourth rows of the table, $\triangle D = 2[-22.95 - (-25.08)] = 4.26$, provides only weak evidence of a difference between ulcer sites, possibly due to the lack of statistical power (p-value = 0.04 from the distribution $\chi^2(1)$).

The fit of the model with all three two-way interactions is shown in Table 9.12. The goodness of fit statistics for this table are $X^2 = 6.49$ and $D = 6.28$ which suggest that the model is not particularly good (compared with the $\chi^2(1)$ distribution) even though $p = 7$ parameters have been used to describe $N = 8$ data points.

Table 9.11 *Results of log-linear modelling of data in Table 9.7.*

Terms in model	d.f.*	log-likelihood**
$GD + CC + GD \times CC$	4	-83.16
$GD + CC + GD \times CC + AP$	3	-30.70
$GD + CC + GD \times CC + AP + AP \times CC$	2	-25.08
$GD + CC + GD \times CC + AP + AP \times CC$ $+ AP \times GD$	1	-22.95

*d.f. denotes degrees of freedom = number of observations (8) minus number of parameters;
**maximum value of the log-likelihood function.

Table 9.12 *Comparison of observed frequencies and expected frequencies obtained from the log-linear model with all two-way interaction terms for the data in Table 9.7; expected frequencies in brackets.*

	Aspirin use		
	Non-user	User	Total
Gastric ulcer			
Controls	62 (58.53)	6 (9.47)	68
Cases	39 (42.47)	25 (21.53)	64
Duodenal ulcer			
Controls	53 (56.47)	8 (4.53)	61
Cases	49 (45.53)	8 (11.47)	57

9.8 Remarks

Two issues relevant to the analysis of a count data have not yet been discussed in this chapter.

First, **overdispersion** occurs when $\text{var}(Y_i)$ is greater than $\text{E}(Y_i)$, although $\text{var}(Y_i) = \text{E}(Y_i)$ for the Poisson distribution. The **negative binomial distribution** provides an alternative model with $\text{var}(Y_i) = \phi\, \text{E}(Y_i)$, where $\phi > 1$ is a parameter that can be estimated. Overdispersion can be due to lack of independence between the observations, in which case the methods described in Chapter 11 for correlated data can be used.

Second, contingency tables may include cells which cannot have any observations (e.g., male hysterectomy cases). This phenomenon, termed **structural zeros**, may not be easily incorporated in Poisson regression unless the parameters can be specified to accommodate the situation. Alternative approaches are discussed by Agresti (1990).

9.9 Exercises

9.1 Let $Y_1, ..., Y_N$ be independent random variables with $Y_i \sim$ Poisson (μ_i) and $\log \mu_i = \beta_1 + \sum_{j=2}^{J} x_{ij}\beta_j, \quad i = 1, ..., N$.

(a) Show that the score statistic for β_1 is $U_1 = \sum_{i=1}^{N}(Y_i - \mu_i)$.

(c) Hence show that for maximum likelihood estimates $\widehat{\mu}_i$, $\sum\widehat{\mu}_i = \sum y_i$.

(d) Deduce that the expression for the deviance in (9.6) simplifies to (9.7) in this case.

9.2 The data in Table 9.13 are numbers of insurance policies, n, and numbers of claims, y, for cars in various insurance categories, CAR, tabulated by age of policy holder, AGE, and district where the policy holder lived ($DIST = 1$, for London and other major cities and $DIST = 0$, otherwise). The table is derived from the $CLAIMS$ data set in Aitkin et al. (1989) obtained from a paper by Baxter, Coutts and Ross (1980).

(a) Calculate the rate of claims y/n for each category and plot the rates by AGE, CAR and $DIST$ to get an idea of the main effects of these factors.

(b) Use Poisson regression to estimate the main effects (each treated as categorical and modelled using indicator variables) and interaction terms.

(c) Based on the modelling in (b), Aitkin et al. (1989) determined that all the interactions were unimportant and decided that AGE and CAR could be treated as though they were continuous variables. Fit a model incorporating these features and compare it with the best model obtained in (b). What conclusions do you reach?

9.3(a) Using a conventional chi-squared test and an appropriate log-linear model, test the hypothesis that the distribution of responses is the same for the placebo and vaccine groups for the flu vaccine trial data in Table 9.6.

(b) For the model corresponding to the hypothesis of homogeneity of response distributions, calculate the fitted values, the Pearson and deviance residuals and the goodness of fit statistics X^2 and D. Which of the cells of the table contribute most to X^2 (or D)? Explain and interpret these results.

(c) Re-analyze these data using ordinal logistic regression to estimate cut-points for a latent continuous response variable and to estimate a location shift between the two treatment groups. Sketch a rough diagram to illustrate the model which forms the conceptual base for this analysis (see Exercise 8.4).

9.4 For a 2×2 contingency table, the maximal log-linear model can be written as

$$\begin{aligned} \eta_{11} &= \mu + \alpha + \beta + (\alpha\beta), \quad \eta_{12} = \mu + \alpha - \beta - (\alpha\beta), \\ \eta_{21} &= \mu - \alpha + \beta - (\alpha\beta), \quad \eta_{22} = \mu - \alpha - \beta + (\alpha\beta), \end{aligned}$$

where $\eta_{jk} = \log\mathrm{E}(Y_{jk}) = \log(n\theta_{jk})$ and $n = \sum\sum Y_{jk}$.

Table 9.13 *Car insurance claims: based on the CLAIMS data set reported by Aitkin et al. (1989).*

		DIST = 0		*DIST* = 1	
CAR	*AGE*	*y*	*n*	*y*	*n*
1	1	65	317	2	20
1	2	65	476	5	33
1	3	52	486	4	40
1	4	310	3259	36	316
2	1	98	486	7	31
2	2	159	1004	10	81
2	3	175	1355	22	122
2	4	877	7660	102	724
3	1	41	223	5	18
3	2	117	539	7	39
3	3	137	697	16	68
3	4	477	3442	63	344
4	1	11	40	0	3
4	2	35	148	6	16
4	3	39	214	8	25
4	4	167	1019	33	114

Show that the interaction term $(\alpha\beta)$ is given by

$$(\alpha\beta) = \tfrac{1}{4}\log\phi$$

where ϕ is the **odds ratio** $(\theta_{11}\ \theta_{22})/(\theta_{12}\ \theta_{21})$, and hence that $\phi = 1$ corresponds to no interaction.

9.5 Use log-linear models to examine the housing satisfaction data in Table 8.5. The numbers of people surveyed in each type of housing can be regarded as fixed.

(a) First analyze the associations between level of satisfaction (treated as a nominal categorical variable) and contact with other residents, separately for each type of housing.

(b) Next conduct the analyses in (a) simultaneously for all types of housing.

(c) Compare the results from log-linear modelling with those obtained using nominal or ordinal logistic regression (see Exercise 8.2).

9.6 Consider a $2\times K$ contingency table (Table 9.14) in which the column totals $y_{.k}$ are fixed for $k = 1, ..., K$.

(a) Show that the product multinomial distribution for this table reduces to

$$f(z_1, ..., z_K\,/n_1, ..., n_K) = \sum_{k=1}^{K} \binom{n_k}{z_k} \pi_k^{z_k}(1-\pi_k)^{n_k - z_k}$$

Table 9.14 *Contingency table with 2 rows and K columns.*

	1	...	k	...	K
Success	y_{11}		y_{1k}		y_{1K}
Failure	y_{21}		y_{2k}		y_{2K}
Total	$y_{.1}$		$y_{.k}$		$y_{.K}$

where $n_k = y_{.k}, z_k = y_{1k}, n_k - z_k = y_{2k}, \pi_k = \theta_{1k}$ and $1 - \pi_k = \theta_{2k}$, for $k = 1, ..., K$. This is the **product binomial distribution** and is the joint distribution for Table 7.1 (with appropriate changes in notation).

(b) Show that the log-linear model with

$$\eta_{1k} = \log \mathrm{E}\,(Z_k) = \mathbf{x}_{1k}^T \boldsymbol{\beta}$$

and

$$\eta_{2k} = \log \mathrm{E}\,(n_k - Z_k) = \mathbf{x}_{2k}^T \boldsymbol{\beta}$$

is equivalent to the logistic model

$$\log \left(\frac{\pi_k}{1 - \pi_k} \right) = \mathbf{x}_k^T \boldsymbol{\beta}$$

where $\mathbf{x}_k = \mathbf{x}_{1k} - \mathbf{x}_{2k}$, $k = 1, ..., K$.

(c) Based on (b), analyze the case-control study data on aspirin use and ulcers using logistic regression and compare the results with those obtained using log-linear models.

10

Survival Analysis

10.1 Introduction

An important type of data is the time from a well-defined starting point until some event, called 'failure', occurs. In engineering, this may be the time from initial use of a component until it fails to operate properly. In medicine, it may be the time from when a patient is diagnosed with a disease until he or she dies. Analysis of these data focuses on summarizing the main features of the distribution, such as median or other percentiles of time to failure, and on examining the effects of explanatory variables. Data on times until failure, or more optimistically, duration of survival or **survival times,** have two important features:

(a) the times are non-negative and typically have skewed distributions with long tails;

(b) some of the subjects may survive beyond the study period so that their actual failure times may not be known; in this case, and other cases where the failure times are not known completely, the data are said to be **censored**.

Examples of various forms of censoring are shown in Figure 10.1. The horizontal lines represent the survival times of subjects. T_O and T_C are the beginning and end of the study period, respectively. D denotes 'death' or 'failure' and A denotes 'alive at the end of the study'. L indicates that the subject was known to be alive at the time shown but then became lost to the study and so the subsequent life course is unknown.

For subjects 1 and 2, the entire survival period (e.g., from diagnosis until death, or from installation of a machine until failure) occurred within the study period. For subject 3, 'death' occurred after the end of the study so that only the solid part of the line is recorded and the time is said to be **right censored** at time T_C. For subject 4, the observed survival time was right censored due to loss of follow up at time T_L. For subject 5, the survival time commenced before the study began so the period before T_O (i.e., the dotted line) is not recorded and the recorded survival time is said to be **left censored** at time T_O.

The analysis of survival time data is the topic of numerous books and papers. Procedures to implement the calculations are available in most statistical programs. In this book, only continuous scale survival time data are considered. Furthermore only parametric models are considered; that is, models requiring the specification of a probability distribution for the survival times. In particular, this means that one of the best known forms of survival analysis, the **Cox proportional hazards model** (Cox, 1972), is not considered because it is a **semi-parametric model** in which dependence on the explanatory variables is modelled explicitly but no specific probability distribution is

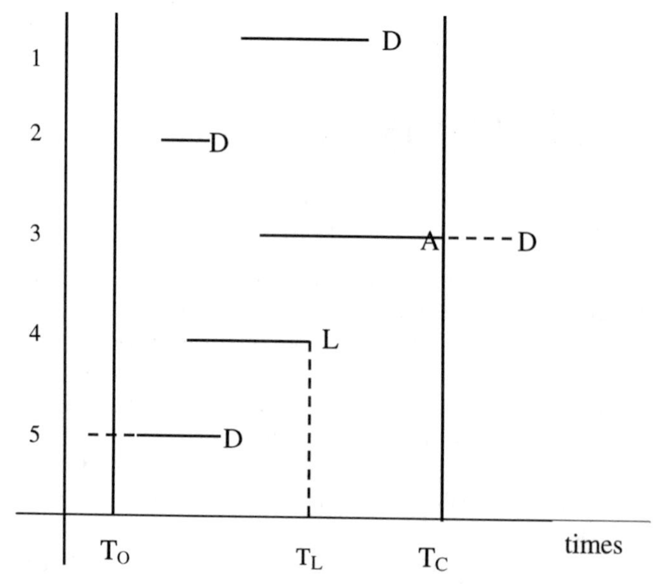

Figure 10.1 *Diagram of types of censoring for survival times.*

assumed for the survival times. An advantage of parametric models, compared to the Cox proportional hazards model, is that inferences are usually more precise and there is a wider range of models with which to describe the data, including **accelerated failure time models** (Wei, 1992). Important topics not considered here include time dependent explanatory variables (Kalbfleisch and Prentice, 1980) and discrete survival time models (Fleming and Harrington, 1991). Fairly recent books that describe the analysis of survival data in detail include Collett (1994), Lee (1992), Cox and Oakes (1984) and Crowder et al. (1991).

The next section explains various functions of the probability distribution of survival times which are useful for model specification. This is followed by descriptions of the two distributions most commonly used for survival data – the exponential and Weibull distributions.

Estimation and inference for survival data are complicated by the presence of censored survival times. The likelihood function contains two components, one involving the uncensored survival times and the other making as much use as possible of information about the survival times which are censored. For several of the more commonly used probability distributions the requirements for generalized linear models are not fully met. Nevertheless, estimation based on the Newton-Raphson method for maximizing the likelihood function, described in Chapter 4, and the inference methods described in Chapter 5 all apply quite well, at least for large sample sizes.

The methods discussed in this chapter are illustrated using a small data set so that the calculations are relatively easy, even though the asymptotic properties of the methods apply only approximately.

10.2 Survivor functions and hazard functions

Let the random variable Y denote the survival time and let $f(y)$ denote its probability density function. Then the probability of failure before a specific time y is given by the cumulative probability distribution

$$F(y) = \Pr(Y < y) = \int_0^y f(t)dt.$$

The **survivor function** is the probability of survival beyond time y. It is given by

$$S(y) = \Pr(Y \geq y) = 1 - F(y). \tag{10.1}$$

The **hazard function** is the probability of death in an infinitesimally small time between y and $(y + \delta y)$, given survival up to time y,

$$\begin{aligned} h(y) &= \lim_{\delta y \to 0} \frac{\Pr(y \leqslant Y < y + \delta y \mid Y > y)}{\delta y} \\ &= \lim_{\delta y \to 0} \frac{F(y + \delta y) - F(y)}{\delta y} \times \frac{1}{S(y)}. \end{aligned}$$

But

$$\lim_{\delta y \to 0} \frac{F(y + \delta y) - F(y)}{\delta y} = f(y)$$

by the definition of a derivative. Therefore,

$$h(y) = \frac{f(y)}{S(y)} \tag{10.2}$$

which can also be written as

$$h(y) = -\frac{d}{dy}\{\log[S(y)]\}. \tag{10.3}$$

Hence

$$S(y) = \exp[-H(y)] \text{ where } H(y) = \int_0^y h(t)dt$$

or

$$H(y) = -\log[S(y)]. \tag{10.4}$$

$H(y)$ is called the **cumulative hazard function** or **integrated hazard function.**

The ‘average’ survival time is usually estimated by the median of the distribution. This is preferable to the expected value because of the skewness of the distribution. The **median survival time**, $y(50)$, is given by the solution of the equation $F(y) = \frac{1}{2}$. Other percentiles can be obtained similarly; for example, the pth percentile $y(p)$ is the solution of $F[y(p)] = p/100$ or $S[y(p)] = 1 - (p/100)$. For some distributions these percentiles may be obtained explicitly; for others, the percentiles may need to be calculated from the estimated survivor function (see Section 10.6).

10.2.1 Exponential distribution

The simplest model for a survival time Y is the **exponential distribution** with probability density function

$$f(y;\theta) = \theta e^{-\theta y}, \qquad y \geq 0,\, \theta > 0. \tag{10.5}$$

This is a member of the exponential family of distributions (see Exercise 3.3(b)) and has $\text{E}(Y){=}1/\theta$ and $\text{var}(Y){=}1/\theta^2$ (see Exercise 4.2). The cumulative distribution is

$$F(y;\theta) = \int_0^y \theta e^{-\theta t} dt = 1 - e^{-\theta y}.$$

So the survivor function is

$$S(y;\theta) = e^{-\theta y}, \tag{10.6}$$

the hazard function is

$$h(y;\theta) = \theta$$

and the cumulative hazard function is

$$H(y;\theta) = \theta y.$$

The hazard function does not depend on y so the probability of failure in the time interval $[y, y + \delta y]$ is not related to how long the subject has already survived. This '**lack of memory**' property may be a limitation because, in practice, the probability of failure often increases with time. In such situations an accelerated failure time model, such as the Weibull distribution, may be more appropriate. One way to examine whether data satisfy the constant hazard property is to estimate the cumulative hazard function $H(y)$ (see Section 10.3) and plot it against survival time y. If the plot is nearly linear then the exponential distribution may provide a useful model for the data.

The **median survival time** is given by the solution of the equation

$$F(y;\theta) = \frac{1}{2} \quad \text{which is} \quad y(50) = \frac{1}{\theta}\log 2.$$

This is a more appropriate description of the 'average' survival time than $\text{E}(Y) = 1/\theta$ because of the skewness of the exponential distribution.

10.2.2 Proportional hazards models

For an exponential distribution, the dependence of Y on explanatory variables could be modelled as $\mathrm{E}(Y) = \mathbf{x}^T\boldsymbol{\beta}$. In this case the identity link function would be used. To ensure that $\theta > 0$, however, it is more common to use

$$\theta = e^{\mathbf{x}^T\boldsymbol{\beta}}.$$

In this case the hazard function has the multiplicative form

$$h(y;\boldsymbol{\beta}) = \theta = e^{\mathbf{x}^T\boldsymbol{\beta}} = \exp(\sum_{i=1}^{p} x_i\beta_i).$$

For a binary explanatory variable with values $x_k = 0$ if the exposure is absent and $x_k = 1$ if the exposure is present, the **hazard ratio** or **relative hazard** for presence vs. absence of exposure is

$$\frac{h_1(y;\boldsymbol{\beta})}{h_0(y;\boldsymbol{\beta})} = e^{\beta_k} \tag{10.7}$$

provided that $\sum_{i\neq k} x_i\beta_i$ is constant. A one-unit change in a continuous explanatory variable x_k will also result in the hazard ratio given in (10.7).

More generally, models of the form

$$h_1(y) = h_0(y)e^{\mathbf{x}^T\boldsymbol{\beta}} \tag{10.8}$$

are called **proportional hazards models** and $h_0(y)$, which is the hazard function corresponding to the reference levels for all the explanatory variables, is called the **baseline hazard**.

For proportional hazards models, the cumulative hazard function is given by

$$H_1(y) = \int_0^y h_1(t)dt = \int_0^y h_0(t)e^{\mathbf{x}^T\boldsymbol{\beta}}dt = H_0(y)e^{\mathbf{x}^T\boldsymbol{\beta}}$$

so

$$\log H_1(y) = \log H_0(y) + \sum_{i=1}^{p} x_i\beta_i.$$

Therefore, for two groups of subjects which differ only with respect to the presence (denoted by P) or absence (denoted by A) of some exposure, from (10.7)

$$\log H_P(y) = \log H_A(y) + \beta_k \tag{10.9}$$

so the **log cumulative hazard functions** differ by a constant.

10.2.3 Weibull distribution

Another commonly used model for survival times is the Weibull distribution which has the probability density function

$$f(y;\lambda,\theta)=\frac{\lambda y^{\lambda-1}}{\theta^{\lambda}}\exp\left[-(\frac{y}{\theta})^{\lambda}\right],\qquad y\geq 0,\quad \lambda>0,\quad \theta>0$$

(see Example 4.2). The parameters λ and θ determine the shape of the distribution and the scale, respectively. To simplify some of the notation, it is convenient to reparameterize the distribution using $\theta^{-\lambda}=\phi$. Then the probability density function is

$$f(y;\lambda,\phi)=\lambda\phi y^{\lambda-1}\exp\left(-\phi y^{\lambda}\right). \tag{10.10}$$

The exponential distribution is a special case of the Weibull distribution with $\lambda=1$.

The survivor function for the Weibull distribution is

$$\begin{aligned} S\left(y;\lambda,\phi\right) &= \int_{y}^{\infty}\lambda\phi u^{\lambda-1}\exp\left(-\phi u^{\lambda}\right)du \\ &= \exp\left(-\phi y^{\lambda}\right), \end{aligned} \tag{10.11}$$

the hazard function is

$$h\left(y;\lambda,\phi\right)=\lambda\phi y^{\lambda-1} \tag{10.12}$$

and the cumulative hazard function is

$$H\left(y;\lambda,\phi\right)=\phi y^{\lambda}.$$

The hazard function depends on y and with suitable values of λ it can increase or decrease with increasing survival time. Thus, the Weibull distribution yields **accelerated failure time** models. The appropriateness of this feature for modelling a particular data set can be assessed using

$$\begin{aligned} \log H(y) &= \log\phi+\lambda\log y \\ &= \log[-\log S(y)]. \end{aligned} \tag{10.13}$$

The empirical survivor function $\widehat{S}(y)$ can be used to plot $\log[-\log\widehat{S}(y)]$ (or $\widehat{S}(y)$ can be plotted on the complementary log-log scale) against the logarithm of the survival times. For the Weibull (or exponential) distribution the points should lie approximately on a straight line. This technique is illustrated in Section 10.3.

It can be shown that the expected value of the survival time Y is

$$\begin{aligned} \mathrm{E}\left(Y\right) &= \int_{0}^{\infty}\lambda\phi y^{\lambda}\exp\left(-\phi y^{\lambda}\right)dy \\ &= \phi^{-1/\lambda}\Gamma\left(1+1/\lambda\right) \end{aligned}$$

where $\Gamma(u) = \int_0^\infty s^{u-1}e^{-s}ds$. Also the median, given by the solution of

$$S(y;\lambda,\phi) = \frac{1}{2},$$

is

$$y(50) = \phi^{-1/\lambda}(\log 2)^{1/\lambda}.$$

These statistics suggest that the relationship between Y and explanatory variables should be modelled in terms of ϕ and it should be multiplicative. In particular, if

$$\phi = \alpha e^{\mathbf{x}^T\boldsymbol{\beta}}$$

then the hazard function (10.12) becomes

$$h(y;\lambda,\phi) = \lambda\alpha y^{\lambda-1}e^{\mathbf{x}^T\boldsymbol{\beta}}. \tag{10.14}$$

If $h_0(y)$ is the baseline hazard function corresponding to reference levels of all the explanatory variables, then

$$h(y) = h_0(y)e^{\mathbf{x}^T\boldsymbol{\beta}}$$

which is a proportional hazards model.

In fact, the Weibull distribution is the only distribution for survival time data that has the properties of accelerated failure times and proportional hazards; see Exercises 10.3 and 10.4 and Cox and Oakes (1984).

10.3 Empirical survivor function

The cumulative hazard function $H(y)$ is an important tool for examining how well a particular distribution describes a set of survival time data. For example, for the exponential distribution, $H(y) = \theta y$ is a linear function of time (see Section 10.2.1) and this can be assessed from the data.

The empirical survivor function, an estimate of the probability of survival beyond time y, is given by

$$\widetilde{S}(y) = \frac{\text{number of subjects with survival times } \geqslant y}{\text{total number of subjects}}.$$

The most common way to calculate this function is to use the **Kaplan Meier estimate**, which is also called the **product limit estimate**. It is calculated by first arranging the observed survival times in order of increasing magnitude $y_{(1)} \leqslant y_{(2)} \leqslant \ldots \leqslant y_{(k)}$. Let n_j denote the number of subjects who are alive just before time $y_{(j)}$ and let d_j denote the number of deaths that occur at time $y_{(j)}$ (or, strictly within a small time interval from $y_{(j)} - \delta$ to $y_{(j)}$). Then the estimated probability of survival past $y_{(j)}$ is $(n_j - d_j)/n_j$. Assuming that the times $y_{(j)}$ are independent, the Kaplan Meier estimate of the survivor

Table 10.1 *Remission times of leukemia patients; data from Gehan (1965).*

Controls										
1	1	2	2	3	4	4	5	5	8	8
8	8	11	11	12	12	15	17	22	23	
Treatment										
6	6	6	6*	7	9*	10	10*	11*	13	16
17*	19*	20*	22	23	25*	32*	32*	34*	35*	
* indicates censoring										

Table 10.2 *Calculation of Kaplan Meier estimate of the survivor function for the treatment group for the data in Table 10.1.*

Time y_j	No. n_j alive just before time y_j	No. d_j deaths at time y_j	$\widehat{S}(y) = \prod\left(\frac{n_j-d_j}{n_j}\right)$
0-<6	21	0	1
6-<7	21	3	0.857
7-<10	17	1	0.807
10-<13	15	1	0.753
13-<16	12	1	0.690
16-<22	11	1	0.627
22-<23	7	1	0.538
≥23	6	1	0.448

function at time y is

$$\widehat{S}(y) = \prod_{j=1}^{k}\left(\frac{n_j - d_j}{n_j}\right)$$

for y between times $y_{(j)}$ and $y_{(j+1)}$.

10.3.1 *Example: Remission times*

The calculation of $\widehat{S}(y)$ is illustrated using an old data set of times to remission of leukemia patients (Gehan, 1965). There are two groups each of $n = 21$ patients. In the control group who were treated with a placebo there was no censoring, whereas in the active treatment group, who were given 6 mercaptopurine, more than half of the observations were censored. The data for both groups are given in Table 10.1. Details of the calculation of $\widehat{S}(y)$ for the treatment group are shown in Table 10.2.

Figure 10.2 shows dot plots of the uncensored times (dots) and censored times (squares) for each group. Due to the high level of censoring in the treatment group, the distributions are not really comparable. Nevertheless,

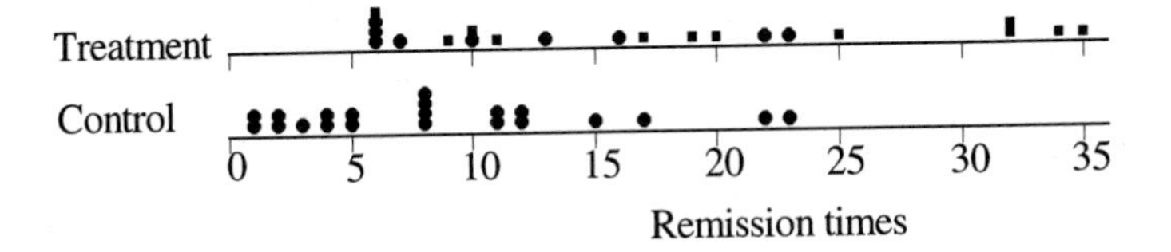

Figure 10.2 *Dot plots of remission time data in Table 10.1: dots represent uncensored times and squares represent censored times.*

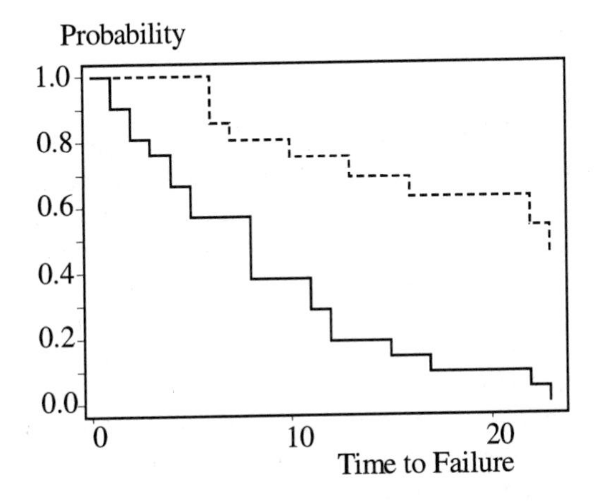

Figure 10.3 *Empirical survivor functions (Kaplan Meier estimates) for data in Table 10.1: the solid line represents the control group and the dotted line represents the treatment group.*

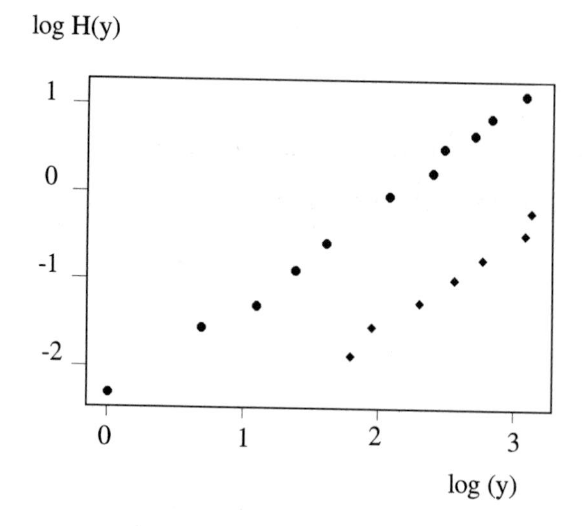

Figure 10.4 *Log cumulative hazard function plotted against log of remission time for data in Table 10.1; dots represent the control group and diamonds represent the treatment group.*

the plots show the skewed distributions and suggest that survival times were longer in the treatment group. Figure 10.3 shows the Kaplan Meier estimates of the survivor functions for the two groups. The solid line represents the control group and the dotted line represents the treatment group. Survival was obviously better in the treatment group. Figure 10.4 shows the logarithm of the cumulative hazard function plotted against log y. The two lines are fairly straight which suggests that the Weibull distribution is appropriate, from (10.13). Furthermore, the lines are parallel which suggests that the proportional hazards model is appropriate, from (10.9). The slopes of the lines are near unity which suggests that the simpler exponential distribution may provide as good a model as the Weibull distribution. The distance between the lines is about 1.4 which indicates that the hazard ratio is about $\exp(1.4) \simeq 4$, from (10.9).

10.4 Estimation

For the jth subject, the data recorded are: y_j the survival time; δ_j a censoring indicator with $\delta_j = 1$ if the survival time is uncensored and $\delta_j = 0$ if it is censored; and $\mathbf{x}_j$ a vector of explanatory variables. Let $y_1, \ldots, y_r$ denote the uncensored observations and $y_{r+1}, \ldots, y_n$ denote the censored ones. The contribution of the uncensored variables to the likelihood function is

$$\prod_{j=1}^{r} f(y_j).$$

For a censored variable we know the survival time Y is at least y_j $(r+1 \leqslant j \leqslant n)$ and the probability of this is $\Pr(Y \geqslant y_j) = S(y_j)$, so the contribution of the censored variables to the likelihood function is

$$\prod_{j=r+1}^{n} S(y_j).$$

The full likelihood is

$$L = \prod_{j=1}^{n} f(y_j)^{\delta_j} S(y_j)^{1-\delta_j} \tag{10.15}$$

so the log-likelihood function is

$$\begin{aligned} l &= \sum_{j=1}^{n} [\delta_j \log f(y_j) + (1-\delta_j) \log S(y_j)] \\ &= \sum_{j=1}^{n} [\delta_j \log h(y_j) + \log S(y_j)] \end{aligned} \tag{10.16}$$

from Equation (10.2). These functions depend on the parameters of the probability distributions and the parameters in the linear component $\mathbf{x}^T\boldsymbol{\beta}$.

The parameters can be estimated using the methods described in Chapter 4. Usually numerical maximization of the log-likelihood function, based on the Newton-Raphson method, is employed. The inverse of the information matrix which is used in the iterative procedure provides an asymptotic estimate of the variance-covariance matrix of the parameter estimates.

The main difference between the parametric models for survival data described in this book and the commonly used Cox proportional hazards regression model is in the function (10.15). For the Cox model, the functions f and S are not fully specified; for more details, see Collett (1994), for example.

10.4.1 Example: simple exponential model

Suppose we have survival time data with censoring but no explanatory variables and that we believe that the exponential distribution is a suitable model. Then the likelihood function is $L(\theta; \mathbf{y}) = \prod_{j=1}^{n} (\theta e^{-\theta y_j})^{\delta_j} (e^{-\theta y_j})^{1-\delta_j}$ from equations (10.5), (10.6) and (10.15). The log-likelihood function is

$$l(\theta; \mathbf{y}) = \sum_{j=1}^{n} \delta_j \log \theta + \sum_{j=1}^{n} [\delta_j (-\theta y_j) + (1-\delta_j)(-\theta y_j)]. \tag{10.17}$$

As there are r uncensored observations (with $\delta_j = 1$) and $(n-r)$ censored observations (with $\delta_j = 0$), equation (10.17) can be simplified to

$$l(\theta; \mathbf{y}) = r \log \theta - \theta \sum_{j=1}^{n} y_j.$$

The solution of the equation

$$U = \frac{dl(\theta; \mathbf{y})}{d\theta} = \frac{r}{\theta} - \sum y_j = 0$$

gives the maximum likelihood estimator

$$\widehat{\theta} = \frac{r}{\sum Y_j}.$$

If there were no censored observations then $r = n$ and $1/\widehat{\theta}$ is just the mean of the survival times, as might be expected because $\mathrm{E}(Y)=1/\theta$.

The variance of $\widehat{\theta}$ can be obtained from the information

$$\mathrm{var}(\widehat{\theta}) = \frac{1}{\mathfrak{I}} = \frac{-1}{\mathrm{E}(U')}$$

where

$$U' = \frac{d^2 l}{d\theta^2} = \frac{-r}{\theta^2}.$$

So $\mathrm{var}(\widehat{\theta}) = \theta^2/r$ which can be estimated by $\widehat{\theta}^2/r$. Therefore, for example, an approximate 95% confidence interval for θ is $\widehat{\theta} \pm 1.96\, \widehat{\theta}/\sqrt{r}$.

10.4.2 Example: Weibull proportional hazards model

If the data for subject j are $\{y_j, \delta_j$ and $\mathbf{x}_j\}$ and the Weibull distribution is thought to provide a suitable model (for example, based on initial exploratory analysis) then the log-likelihood function is

$$l = \sum_{j=1}^{n} \left[\delta_j \log(\lambda \alpha y_j^{\lambda-1} e^{\mathbf{x}_j^T \boldsymbol{\beta}}) - (\alpha y_j^{\lambda} e^{\mathbf{x}_j^T \boldsymbol{\beta}}) \right]$$

from equations (10.14) and (10.16). This function can be maximized numerically to obtain estimates $\widehat{\lambda}, \widehat{\alpha}$ and $\widehat{\boldsymbol{\beta}}$.

10.5 Inference

The Newton-Raphson iteration procedure used to obtain maximum likelihood estimates also produces the information matrix $\mathfrak{I}$ which can be inverted to give the approximate variance-covariance matrix for the estimators. Hence inferences for any parameter θ can be based on the maximum likelihood estimator $\widehat{\theta}$ and the standard error, $s.e.(\widehat{\theta})$, obtained by taking the square root of the relevant element of the diagonal of $\mathfrak{I}^{-1}$. Then the Wald statistic $(\widehat{\theta} - \theta)/s.e.(\widehat{\theta})$

can be used to test hypotheses about θ or to calculate approximate confidence limits for θ assuming that the statistic has the standard Normal distribution $N(0,1)$ (see Section 5.4).

For the Weibull and exponential distributions, the maximum value of the log-likelihood function can be calculated by substituting the maximum likelihood estimates of the parameters, denoted by the vector $\widehat{\boldsymbol{\theta}}$, into the expression in (10.16) to obtain $l(\widehat{\boldsymbol{\theta}};\mathbf{y})$. For censored data, the statistic $-2l(\widehat{\boldsymbol{\theta}};\mathbf{y})$ may not have a chi-squared distribution, even approximately. For nested models M_1, with p parameters and maximum value $\widehat{l}_1$ of the log-likelihood function, and M_0, with $q<p$ parameters and maximum value $\widehat{l}_0$ of the log-likelihood function, the difference

$$D = 2(\widehat{l}_1 - \widehat{l}_0)$$

will approximately have a chi-squared distribution with $p-q$ degrees of freedom if both models fit well. The statistic D, which is analogous to the **deviance**, provides another method for testing hypotheses (see Section 5.7).

10.6 Model checking

To assess the adequacy of a model it is necessary to check assumptions, such as the proportional hazards and accelerated failure time properties, in addition to looking for patterns in the residuals (see Section 2.3.4) and examining influential observations using statistics analogous to those for multiple linear regression (see Section 6.2.7).

The empirical survivor function $\widehat{S}(y)$ described in Section 10.3 can be used to examine the appropriateness of the probability model. For example, for the exponential distribution, the plot of $-\log[\widehat{S}(y)]$ against y should be approximately linear from (10.6). More generally, for the Weibull distribution, the plot of the log cumulative hazard function $\log[-\log\widehat{S}(y)]$ against $\log y$ should be linear, from (10.13). If the plot shows curvature then some alternative model such as the log-logistic distribution may be better (see Exercise 10.2).

The general proportional hazards model given in (10.8) is

$$h(y) = h_0(y)e^{\mathbf{x}^T\boldsymbol{\beta}}$$

where h_0 is the baseline hazard. Consider a binary explanatory variable x_k with values $x_k=0$ if a characteristic is absent and $x_k=1$ if it is present. The log-cumulative hazard functions are related by

$$\log H_P = \log H_A + \beta_k;$$

see (10.9). Therefore if the empirical hazard functions $\widehat{S}(y)$ are calculated separately for subjects with and without the characteristic and the log-cumulative hazard functions $\log[-\log\widehat{S}(y)]$ are plotted against $\log y$, the lines should have the same slope but be separated by a distance β_k.

More generally, parallel lines for the plots of the log cumulative hazard functions provide support for the proportional hazards assumption. For a fairly small number of categorical explanatory variables, the proportional hazards assumption can be examined in this way. If the lines are not parallel this may suggest that there are interaction effects among the explanatory variables. If the lines are curved but still parallel this supports the proportional hazards assumption but suggests that the accelerated failure time model is inadequate. For more complex situations it may be necessary to rely on general diagnostics based on residuals, although these are not specific for investigating the proportional hazards property.

The simplest residuals for survival time data are the **Cox-Snell residuals**. If the survival time for subject j is uncensored then the Cox-Snell residual is

$$r_{Cj} = \widehat{H}_j(y_j) = -\log[\widehat{S}_j(y_j)] \tag{10.18}$$

where $\widehat{H}_j$ and $\widehat{S}_j$ are the estimated cumulative hazard and survivor functions for subject j at time y_j. For proportional hazards models, (10.18) can be written as

$$r_{Cj} = \exp(\mathbf{x}_j^T\widehat{\boldsymbol{\beta}})\widehat{H}_0(y_j)$$

where $\widehat{H}_0(y_j)$ is the baseline hazard function evaluated at y_j.

It can be shown that if the model fits the data well then these residuals have an exponential distribution with a parameter of one. In particular, their mean and variance should be approximately equal to one.

For censored observations, r_{Cj} will be too small and various modifications have been proposed of the form

$$r'_{Cj} = Y_i = \begin{cases} r_{Cj} \text{ for uncensored observations} \\ r_{Cj} + \Delta \text{ for censored observations} \end{cases}$$

where $\Delta = 1$ or $\Delta = \log 2$ (Crowley and Hu, 1977). The distribution of the r'_{Cj}'s can be compared with the exponential distribution with unit mean using exponential probability plots (analogous Normal probability plots) which are available in various statistical software. An exponential probability plot of the residuals r'_{Cj} may be used to identify outliers and systematic departures from the assumed distribution.

Martingale residuals provide an alternative approach. For the jth subject the martingale residual is

$$r_{Mj} = \delta_j - r_{Cj}$$

where $\delta_j = 1$ if the survival time is uncensored and $\delta_j = 0$ if it is censored. These residuals have an expected value of zero but a negatively skewed distribution.

Deviance residuals (which are somewhat misnamed because the sum of their squares is not, in fact, equal to the deviance mentioned in Section 10.5) are defined by

$$r_{Dj} = \text{sign}(r_{Mj})\{-2[r_{Mj} + \delta_j \log(r_{Cj})]\}^{\frac{1}{2}}.$$

The r_{Dj}'s are approximately symmetrically distributed about zero and large values may indicate outlying observations.

Table 10.3 *Results of fitting proportional hazards models based on the exponential and Weibull distributions to the data in Table 10.1.*

	Exponential model	Weibull model
Group β_1	1.53 (0.40)	1.27 (0.31)
Intercept β_0	0.63 (0.55)	0.98 (0.43)
Shape λ	1.00*	1.37 (0.20)

* shape parameter is unity for the exponential distribution.

In principle, any of the residuals r'_{Cj}, r_{Mj} or r_{Dj} are suitable for sequence plots against the order in which the survival times were measured, or any other relevant order (to detect lack of independence among the observations) and for plots against explanatory variables that have been included in the model (and those that have not) to detect any systematic patterns which would indicate that the model is not correctly specified. However, the skewness of the distributions of r'_{Cj} and r_{Mj} makes them less useful than r_{Dj}, in practice.

Diagnostics to identify influential observations can be defined for survival time data, by analogy with similar statistics for multiple linear regression and other generalized linear models. For example, for any parameter β_k delta-betas $\Delta_j\beta_k$, one for each subject j, show the effect on the estimate of β_k caused by the omitting the data for subject j from the calculations. Plotting the $\Delta_j\beta_k$'s against the order of the observations or against the survival times y_j may indicate systematic effects or particularly influential observations.

10.7 Example: remission times

Figure 10.4 suggests that a proportional hazards model with a Weibull, or even an exponential, distribution should provide a good model for the remission time data in Table 10.1. The models are

$$\begin{aligned} h(y) &= \exp(\beta_0 + \beta_1 x), \qquad y \sim \text{Exponential}; \\ h(y) &= \lambda y^{\lambda-1} \exp(\beta_0 + \beta_1 x), \qquad y \sim \text{Weibull}, \end{aligned} \tag{10.19}$$

where $x = 0$ for the control group, $x = 1$ for the treatment group and λ is the shape parameter. The results of fitting these models are shown in Table 10.3. The hypothesis that $\lambda = 1$ can be tested either using the Wald statistic from the Weibull model, i.e., $z = (1.37 - 1.00)/0.20 = 1.85$, or from $D = 2(\widehat{l}_W - \widehat{l}_E) = 3.89$ where $\widehat{l}_W$ and $\widehat{l}_E$ are the maximum values of the log-likelihood functions for the Weibull and exponential models respectively (details not shown here). Comparing z with the standard Normal distribution or D with the chi-squared distribution with one degree of freedom provides only weak evidence against the hypothesis. Therefore, we can conclude that the exponential distribution is about as good as the Weibull distribution for modelling the data. Both models suggest that the parameter β_1 is non-zero

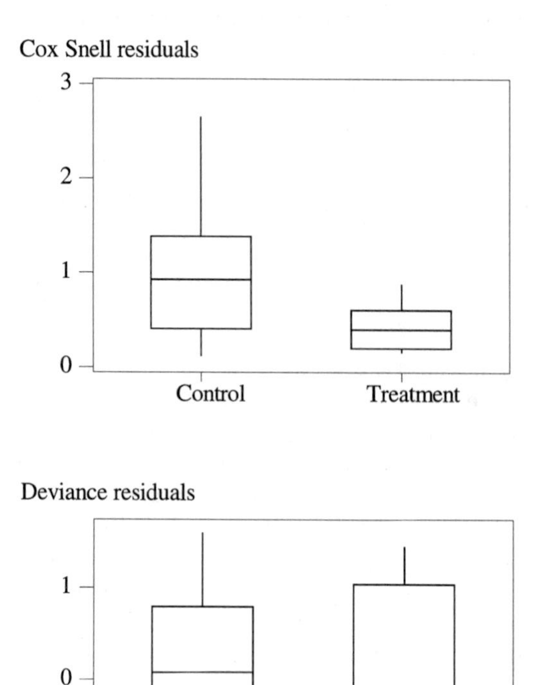

Figure 10.5 *Boxplots of Cox Snell and deviance residuals from the exponential model (10.19) for the data in Table 10.1.*

and the exponential model provides the estimate $\exp(1.53) = 4.62$ for the relative hazard.

Figure 10.5 shows box plots of Cox-Snell and deviance residuals for the exponential model. The skewness of the Cox-Snell residuals and the more symmetric distribution of the deviance residuals is apparent. Additionally, the difference in location between the distributions of the treatment and control groups suggests the model has failed to describe fully the patterns of remission times for the two groups of patients.

Table 10.4 *Leukemia survival times.*

AG positive		AG negative	
Survival time	White blood cell count	Survival time	White blood cell count
65	2.30	56	4.40
156	0.75	65	3.00
100	4.30	17	4.00
134	2.60	7	1.50
16	6.00	16	9.00
108	10.50	22	5.30
121	10.00	3	10.00
4	17.00	4	19.00
39	5.40	2	27.00
143	7.00	3	28.00
56	9.40	8	31.00
26	32.00	4	26.00
22	35.00	3	21.00
1	100.00	30	79.00
1	100.00	4	100.00
5	52.00	43	100.00
65	100.00		

10.8 Exercises

10.1 The data in Table 10.4 are survival times, in weeks, for leukemia patients. There is no censoring. There are two covariates, white blood cell count (WBC) and the results of a test (AG positive and AG negative). The data set is from Feigl and Zelen (1965) and the data for the 17 patients with AG positive test results are described in Exercise 4.2.

(a) Obtain the empirical survivor functions $\widehat{S}(y)$ for each group (AG positive and AG negative), ignoring WBC.

(b) Use suitable plots of the estimates $\widehat{S}(y)$ to select an appropriate probability distribution to model the data.

(c) Use a parametric model to compare the survival times for the two groups, after adjustment for the covariate WBC, which is best transformed to log(WBC).

(d) Check the adequacy of the model using residuals and other diagnostic tests.

(e) Based on this analysis, is AG a useful prognostic indicator?

10.2 The **log-logistic distribution** with the probability density function

$$f(y) = \frac{e^{\theta}\lambda y^{\lambda-1}}{(1+e^{\theta}y^{\lambda})^2}$$

is sometimes used for modelling survival times.

(a) Find the survivor function $S(y)$, the hazard function $h(y)$ and the cumulative hazard function $H(y)$.

(b) Show that the median survival time is $\exp(-\theta/\lambda)$.

(c) Plot the hazard function for $\lambda = 1$ and $\lambda = 5$ with $\theta = -5$, $\theta = -2$ and $\theta = \frac{1}{2}$.

10.3 For **accelerated failure time models** the explanatory variables for subject i, η_i, act multiplicatively on the time variable so that the hazard function for subject i is

$$h_i(y) = \eta_i h_0(\eta_i y)$$

where $h_0(y)$ is the baseline hazard function. Show that the Weibull and log-logistic distributions both have this property but the exponential distribution does not. (Hint: obtain the hazard function for the random variable $T = \eta_i Y$.)

10.4 For **proportional hazards models** the explanatory variables for subject i, η_i, act multiplicatively on the hazard function. If $\eta_i = e^{\mathbf{x}_i^T\boldsymbol{\beta}}$ then the hazard function for subject i is

$$h_i(y) = e^{\mathbf{x}_i^T\boldsymbol{\beta}} h_0(y) \tag{10.20}$$

where $h_0(y)$ is the baseline hazard function.

(a) For the exponential distribution if $h_0 = \theta$ show that if $\theta_i = e^{\mathbf{x}_i^T\boldsymbol{\beta}}\theta$ for the ith subject, then (10.20) is satisfied.

(b) For the Weibull distribution if $h_0 = \lambda\phi y^{\lambda-1}$ show that if $\phi_i = e^{\mathbf{x}_i^T\boldsymbol{\beta}}\phi$ for the ith subject, then (10.20) is satisfied.

(c) For the log-logistic distribution if $h_0 = e^{\theta}\lambda y^{\lambda-1}/(1 + e^{\theta}y^{\lambda})$ show that if $e^{\theta_i} = e^{\theta + \mathbf{x}_i^T\boldsymbol{\beta}}$ for the ith subject, then (10.20) is not satisfied. Hence, or otherwise, deduce that the log-logistic distribution does not have the proportional hazards property.

10.5 As the survivor function $S(y)$ is the probability of surviving beyond time y, the odds of survival past time y is

$$O(y) = \frac{S(y)}{1 - S(y)}.$$

For **proportional odds models** the explanatory variables for subject i, η_i, act multiplicatively on the odds of survival beyond time y

$$O_i = \eta_i O_0$$

where O_0 is the baseline odds.

(a) Find the odds of survival beyond time y for the exponential, Weibull and log-logistic distributions.

(b) Show that only the log-logistic distribution has the proportional odds property.

Table 10.5 *Survival times in months of patients with chronic active hepatitis in a randomized controlled trial of prednisolone versus no treatment; data from Altman and Bland (1998).*

Prednisolone							
2	6	12	54	56**	68	89	96
96	125*	128*	131*	140*	141*	143	145*
146	148*	162*	168	173*	181*		
No treatment							
2	3	4	7	10	22	28	29
32	37	40	41	54	61	63	71
127*	140*	146*	158*	167*	182*		

* indicates censoring, ** indicates loss to follow-up.

(c) For the log-logistic distribution show that the log odds of survival beyond time y is

$$\log O(y) = \log\left[\frac{S(y)}{1-S(y)}\right] = -\theta - \lambda \log y.$$

Therefore if $\log \widehat{O}_i$ (estimated from the empirical survivor function) plotted against $\log y$ is approximately linear, then the log-logistic distribution may provide a suitable model.

(d) From (b) and (c) deduce that for two groups of subjects with explanatory variables η_1 and η_2 plots of $\log \widehat{O}_1$ and $\log \widehat{O}_2$ against $\log y$ should produce approximately parallel straight lines.

10.6 The data in Table 10.5 are survival times, in months, of 44 patients with chronic active hepatitis. They participated in a randomized controlled trial of prednisolone compared with no treatment. There were 22 patients in each group. One patient was lost to follow-up and several in each group were still alive at the end of the trial. The data are from Altman and Bland (1998).

(a) Calculate the empirical survivor functions for each group.

(b) Use suitable plots to investigate the properties of accelerated failure times, proportional hazards and proportional odds, using the results from Exercises 10.3, 10.4 and 10.5 respectively.

(c) Based on the results from (b) fit an appropriate model to the data in Table 10.5 to estimate the relative effect of prednisolone.

11

Clustered and Longitudinal Data

11.1 Introduction

In all the models considered so far the outcomes Y_i, $i = 1, \ldots, n$ are assumed to be independent. There are two common situations where this assumption is implausible. In one situation the outcomes are repeated measurements over time on the same subjects; for example, the weights of the same people when they are 30, 40, 50 and 60 years old. This is an example of **longitudinal data**. Measurements on the same person at different times may be more alike than measurements on different people, because they are affected by persistent characteristics as well as potentially more variable factors; for instance, weight is likely to be related to an adult's genetic makeup and height as well as their eating habits and level of physical activity. For this reason longitudinal data for a group of subjects are likely to exhibit correlation between successive measurements.

The other situation in which data are likely to be correlated is where they are measurements on related subjects; for example, the weights of samples of women aged 40 years selected from specific locations in different countries. In this case the countries are the **primary sampling units** or **clusters** and the women are sub-samples within each primary sampling unit. Women from the same geographic area are likely to be more similar to one another, due to shared socio-economic and environmental conditions, than they are to women from other locations. Any comparison of women's weights between areas that failed to take this within-area correlation into account could produce misleading results. For example, the standard deviation of the mean difference in weights between two areas will be over-estimated if the observations which are correlated are assumed to be independent.

The term **repeated measures** is used to describe both longitudinal and clustered data. In both cases, models that include correlation are needed in order to make valid statistical inferences. There are two approaches to modelling such data.

One approach involves dropping the usual assumption of independence between the outcomes Y_i and modelling the correlation structure explicitly. This method goes under various names such as **repeated measures** (for example, **repeated measures analysis of variance**) and the **generalized estimating equation** approach. The estimation and inference procedures for these models are, in principle, analogous to those for generalized linear models for independent outcomes; although in practice, software may not be readily available to do the calculations.

The alternative approach for modelling repeated measures is based on considering the hierarchical structure of the study design. This is called **mul-**

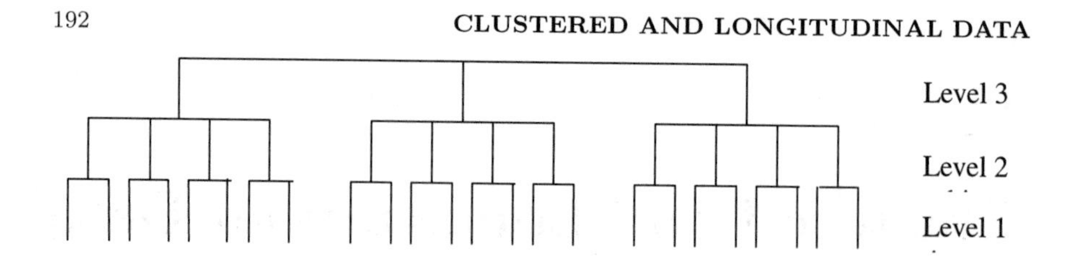

Figure 11.1 *Multilevel study.*

tilevel modelling. For example, suppose there are repeated, longitudinal measurements, level 1, on different subjects, level 2, who were randomized to experimental groups, level 3. This nested structure is illustrated in Figure 11.1 which shows three groups, each of four subjects, on whom measurements are made at two times (for example, before and after some intervention). On each branch, outcomes at the same level are assumed to be independent and the correlation is a result of the multilevel structure (see Section 11.4).

In the next section an example is presented of an experiment with longitudinal outcome measures. Descriptive data analyses are used to explore the study hypothesis and the assumptions that are made in various models which might be used to test the hypothesis.

Repeated measures models for Normal data are described in Section 11.3. In Section 11.4, repeated measures models are described for non-Normal data such as counts and proportions which might be analyzed using Poisson, binomial and other distributions (usually from the exponential family). These sections include details of the relevant estimation and inferential procedures. For repeated measures models, it is necessary to choose a correlation structure likely to reflect the relationships between the observations. Usually the correlation parameters are not of particular interest (i.e., they are nuisance parameters) but they need to be included in the model in order to obtain consistent estimates of those parameters that are of interest and to correctly calculate the standard errors of these estimates.

For multilevel models described in Section 11.5, the effects of levels may be described either by fixed parameters (e.g., for group effects) or random variables (e.g., for subjects randomly allocated to groups). If the linear predictor of the model has both fixed and random effects the term **mixed model** is used. The correlation between observations is due to the random effects. This may make the correlation easier to interpret in multilevel models than in repeated measures models. Also the correlation parameters may be of direct interest. For Normally distributed data, multilevel models are well-established and estimation and model checking procedures are available in most general purpose statistical software. For counts and proportions, although the model specification is conceptually straightforward, there is less software available to fit the models.

In Section 11.6, both repeated measures and multilevel models are fitted to data from the stroke example in Section 11.2. The results are used to illustrate the connections between the various models.

Finally, in Section 11.7, a number of issues that arise in the modelling of clustered and longitudinal data are mentioned. These include methods of exploratory analysis, consequences of using inappropriate models and problems that arise from missing data.

11.2 Example: Recovery from stroke

The data in Table 11.1 are from an experiment to promote the recovery of stroke patients. There were three experimental groups:

A was a new occupational therapy intervention;

B was the existing stroke rehabilitation program conducted in the same hospital where A was conducted;

C was the usual care regime for stroke patients provided in a different hospital.

There were eight patients in each experimental group. The response variable was a measure of functional ability, the Bartel index; higher scores correspond to better outcomes and the maximum score is 100. Each patient was assessed weekly over the eight weeks of the study. The study was conducted by C. Cropper, at the University of Queensland, and the data were obtained from the OzDasl website developed by Gordon Smyth (http://www.maths.uq.edu.au/~gks/data/index.html).

The hypothesis was that the patients in group A would do better than those in groups B or C. Figure 11.2 shows the time course of scores for every patient. Figure 11.3 shows the time course of the average scores for each experimental group. Clearly most patients improved. Also it appears that those in group A recovered best and those in group C did worst (however, people in group C may have started at a lower level).

The scatter plot matrix in Figure 11.4 shows data for all 24 patients at different times. The corresponding Pearson correlation coefficients are given in Table 11.2. These show high positive correlation between measurements made one week apart and decreasing correlation between observations further apart in time.

A **naive analysis**, sometimes called a **pooled analysis**, of these data is to fit an analysis of covariance model in which all 192 observations (for 3 groups $\times$ 8 subjects $\times$ 8 times) are assumed to be independent with

$$\mathrm{E}(Y_{ijk}) = \alpha_i + \beta t_k + e_{ijk} \tag{11.1}$$

where Y_{ijk} is the score at time t_k $(k = 1, \ldots, 8)$ for patient $j (j = 1, \ldots, 8)$ in group i (where $i = 1$ for group A, $i = 2$ for group B and $i = 3$ for group C); α_i is the mean score for group i; β is a common slope parameter; t_k denotes time ($t_k = k$ for week k, $k = 1, \ldots, 8$); and the random error terms e_{ijk} are all assumed to be independent. The null hypothesis H_0: $\alpha_1 = \alpha_2 = \alpha_3$ can be compared with an alternative hypothesis such as H_1: $\alpha_1 > \alpha_2 > \alpha_3$ by fitting models with different group parameters α_i. Figure 11.3 suggests that

Table 11.1 *Functional ability scores measuring recovery from stroke for patients in three experimental groups over 8 weeks of the study.*

		Week							
Subject	Group	1	2	3	4	5	6	7	8
1	A	45	45	45	45	80	80	80	90
2	A	20	25	25	25	30	35	30	50
3	A	50	50	55	70	70	75	90	90
4	A	25	25	35	40	60	60	70	80
5	A	100	100	100	100	100	100	100	100
6	A	20	20	30	50	50	60	85	95
7	A	30	35	35	40	50	60	75	85
8	A	30	35	45	50	55	65	65	70
9	B	40	55	60	70	80	85	90	90
10	B	65	65	70	70	80	80	80	80
11	B	30	30	40	45	65	85	85	85
12	B	25	35	35	35	40	45	45	45
13	B	45	45	80	80	80	80	80	80
14	B	15	15	10	10	10	20	20	20
15	B	35	35	35	45	45	45	50	50
16	B	40	40	40	55	55	55	60	65
17	C	20	20	30	30	30	30	30	30
18	C	35	35	35	40	40	40	40	40
19	C	35	35	35	40	40	40	45	45
20	C	45	65	65	65	80	85	95	100
21	C	45	65	70	90	90	95	95	100
22	C	25	30	30	35	40	40	40	40
23	C	25	25	30	30	30	30	35	40
24	C	15	35	35	35	40	50	65	65

Table 11.2 *Correlation coefficients for the stroke recovery scores in Table 11.1.*

	Week						
	1	2	3	4	5	6	7
Week 2	0.93						
Week 3	0.88	0.92					
Week 4	0.83	0.88	0.95				
Week 5	0.79	0.85	0.91	0.92			
Week 6	0.71	0.79	0.85	0.88	0.97		
Week 7	0.62	0.70	0.77	0.83	0.92	0.96	
Week 8	0.55	0.64	0.70	0.77	0.88	0.93	0.98

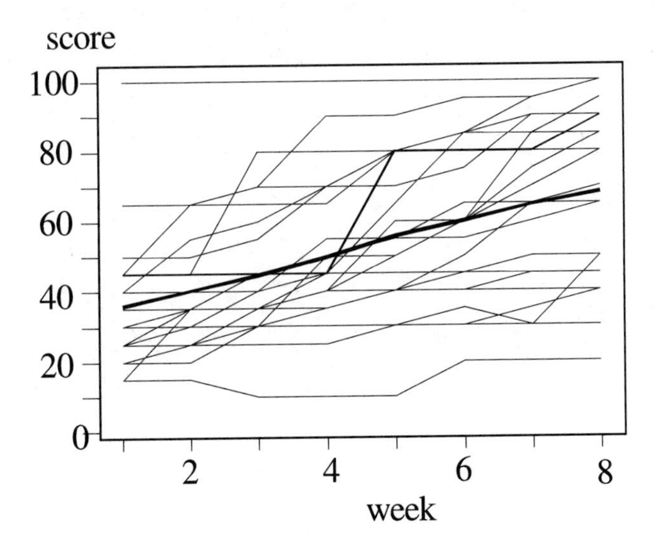

Figure 11.2 *Stroke recovery scores of individual patients.*

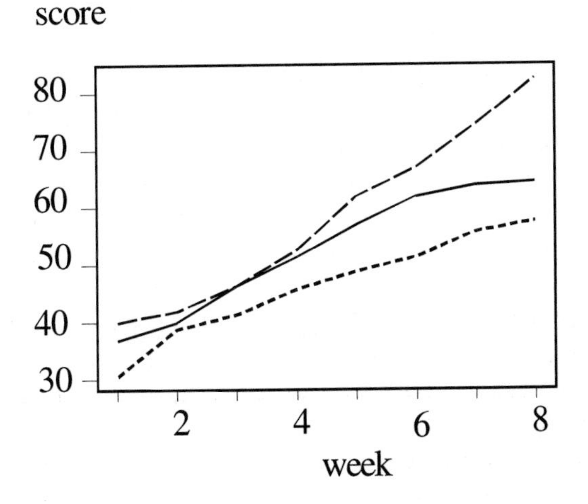

Figure 11.3 *Average stroke recovery scores for groups of patients: long dashed line corresponds to group A; solid line to group B; short dashed line to group C.*

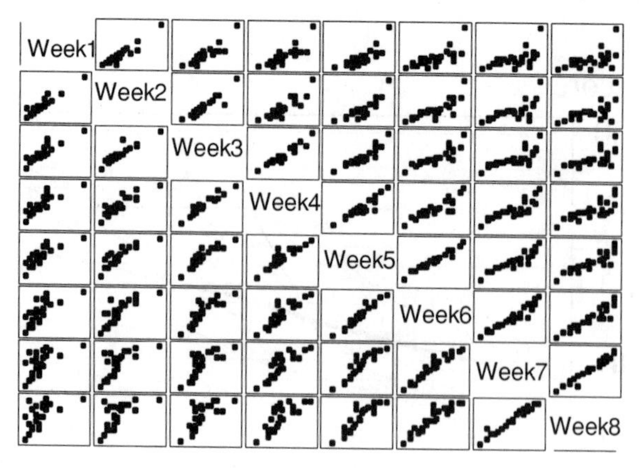

Figure 11.4 *Scatter plot matrix for stroke recovery scores in Table 11.2.*

the slopes may differ between the three groups so the following model was also fitted

$$E(Y_{ijk}) = \alpha_i + \beta_i t_k + e_{ijk}$$

where the slope parameter β_i denotes the rate of recovery for group i. Models (11.1) and (11.2) can be compared to test the hypothesis H_0: $\beta_1 = \beta_2 = \beta_3$ against an alternative hypothesis that the β's differ. Neither of these naive models takes account of the fact that measurements of the same patient at different times are likely to be more similar than measurements of different patients. This is analogous to using an unpaired t-test for paired data (see Exercise 2.2).

Table 11.3 shows the results of fitting these models, which will be compared later with results from more appropriate analyses. Note, however, that for model (11.2) the Wald statistics for $\alpha_2 - \alpha_1$ $(3.348/8.166 = 0.41)$ and for $\alpha_3 - \alpha_1$ $(-0.022/8.166 = -0.003)$ are very small compared to the standard Normal distribution which suggests that the intercepts are not different (i.e., on average the groups started with the same level of functional ability).

A preferable form of **exploratory analysis**, sometimes called **data reduction** or **data summary**, consists of summarizing the response profiles for each subject by a small number of descriptive statistics based on assuming that measurements on the same subject are independent. For the stroke data, appropriate summary statistics are the intercept and slope of the individual regression lines. Other examples of summary statistics that may be appropriate in particular situations include peak values, areas under curves,

Table 11.3 *Results of naive analyses of stroke recovery scores in Table 11.1, assuming all the data are independent and using models (11.1) and (11.2).*

Parameter	Estimate	Standard error
Model (11.1)		
α_1	36.842	3.971
$\alpha_2 - \alpha_1$	-5.625	3.715
$\alpha_3 - \alpha_1$	-12.109	3.715
β	4.764	0.662
Model (11.2)		
α_1	29.821	5.774
$\alpha_2 - \alpha_1$	3.348	8.166
$\alpha_3 - \alpha_1$	-0.022	8.166
β_1	6.324	1.143
$\beta_2 - \beta_1$	-1.994	1.617
$\beta_3 - \beta_1$	-2.686	1.617

or coefficients of quadratic or exponential terms in non-linear growth curves. These subject-specific statistics are used as the data for subsequent analyses.

The intercept and slope estimates and their standard errors for each of the 24 stroke patients are shown in Table 11.4. These results show considerable variability between subjects which should, in principle, be taken into account in any further analyses. Tables 11.5 and 11.6 show analyses comparing intercepts and slopes between the experimental groups, assuming independence between the subjects but ignoring the differences in precision (standard errors) between the estimates. Notice that although the estimates are the same as those for model (11.2) in Table 11.3, the standard errors are (correctly) much larger and the data do not provide much evidence of differences in either the intercepts or the slopes.

Although the analysis of subject specific summary statistics does not require the implausible assumption of independence between observations within subjects, it ignores the random error in the estimates. Ignoring this information can lead to underestimation of effect sizes and underestimation of the overall variation (Fuller, 1987). To avoid these biases, models are needed that better describe the data structure that arises from the study design. Such models are described in the next three sections.

11.3 Repeated measures models for Normal data

Suppose there are N study units or subjects with n_i measurements for subject i (e.g., n_i longitudinal observations for person i or n_i observations for cluster i). Let $\mathbf{y}_i$ denote the vector of responses for subject i and let $\mathbf{y}$ denote the vector of responses for all subjects

Table 11.4 *Estimates of intercepts and slopes (and their standard errors) for each subject in Table 11.1.*

Subject	Intercept (std. error)	Slope (std. error)
1	30.000 (7.289)	7.500 (1.443)
2	15.536 (4.099)	3.214 (0.812)
3	39.821 (3.209)	6.429 (0.636)
4	11.607 (3.387)	8.393 (0.671)
5	100.000 (0.000)	0.000 (0.000)
6	0.893 (5.304)	11.190 (1.050)
7	15.357 (4.669)	7.976 (0.925)
8	25.357 (1.971)	5.893 (0.390)
9	38.571 (3.522)	7.262 (0.698)
10	61.964 (2.236)	2.619 (0.443)
11	14.464 (5.893)	9.702 (1.167)
12	26.071 (2.147)	2.679 (0.425)
13	48.750 (8.927)	5.000 (1.768)
14	10.179 (3.209)	1.071 (0.636)
15	31.250 (1.948)	2.500 (0.386)
16	34.107 (2.809)	3.810 (0.556)
17	21.071 (2.551)	1.429 (0.505)
18	34.107 (1.164)	0.893 (0.231)
19	32.143 (1.164)	1.607 (0.231)
20	42.321 (3.698)	7.262 (0.732)
21	48. 571 (6.140)	7.262 (1.216)
22	24.821 (1.885)	2.262 (0.373)
23	22.321 (1.709)	1.845 (0.339)
24	13.036 (4.492)	6.548 (0.890)

Table 11.5 *Analysis of variance of intercept estimates in Table 11.4.*

Source	d.f.	Mean square	F	p-value
Groups	2	30	0.07	0.94
Error	21	459		
Parameter	Estimate	Std. error		
α_1	29.821	7.572		
$\alpha_2 - \alpha_1$	3.348	10.709		
$\alpha_3 - \alpha_1$	-0.018	10.709		

Table 11.6 *Analysis of variance of slope estimates in Table 11.4.*

Source	d.f.	Mean square	F	p-value
Groups	2	15.56	1.67	0.21
Error	21	9.34		

Parameter	Estimate	Std. error
β_1	6.324	1.080
$\beta_2 - \beta_1$	-1.994	1.528
$\beta_3 - \beta_1$	-2.686	1.528

$$\mathbf{y} = \begin{bmatrix} \mathbf{y}_1 \\ \vdots \\ \mathbf{y}_N \end{bmatrix}, \qquad \text{so } \mathbf{y} \text{ has length } \sum_{i=1}^{N} n_i.$$

A Normal linear model for $\mathbf{y}$ is

$$\mathrm{E}(\mathbf{y}) = \mathbf{X}\boldsymbol{\beta} = \boldsymbol{\mu}; \qquad \mathbf{y} \sim \mathbf{N}(\boldsymbol{\mu}, \mathbf{V}), \tag{11.2}$$

where

$$\mathbf{X} = \begin{bmatrix} \mathbf{X}_1 \\ \mathbf{X}_2 \\ \vdots \\ \mathbf{X}_N \end{bmatrix}, \qquad \boldsymbol{\beta} = \begin{bmatrix} \beta_1 \\ \vdots \\ \beta_p \end{bmatrix},$$

$\mathbf{X}_i$ is the $n_i \times p$ design matrix for subject i and $\boldsymbol{\beta}$ is a parameter vector of length p. The variance-covariance matrix for measurements for subject i is

$$\mathbf{V}_i = \begin{bmatrix} \sigma_{i11} & \sigma_{i12} & \cdots & \sigma_{i1n_i} \\ \sigma_{i21} & \ddots & & \vdots \\ \vdots & & \ddots & \\ \sigma_{in1} & & & \sigma_{in_in_i} \end{bmatrix}$$

and the overall variance-covariance matrix has the block diagonal form

$$\mathbf{V} = \begin{bmatrix} \mathbf{V}_1 & \mathbf{O} & & \mathbf{O} \\ \mathbf{O} & \mathbf{V}_2 & & \mathbf{O} \\ & & \ddots & \\ \mathbf{O} & \mathbf{O} & & \mathbf{V}_N \end{bmatrix}$$

assuming that responses for different subjects are independent (where $\mathbf{O}$ denotes a matrix of zeros). Usually the matrices $\mathbf{V}_i$ are assumed to have the same form for all subjects.

If the elements of $\mathbf{V}$ are known constants then $\boldsymbol{\beta}$ can be estimated from the likelihood function for model (11.3) or by the method of least squares. The maximum likelihood estimator, obtained by solving the score equations

$$\mathbf{U}(\boldsymbol{\beta}) = \frac{\partial l}{\partial \boldsymbol{\beta}} = \mathbf{X}^T\mathbf{V}^{-1}(\mathbf{y} - \mathbf{X}\boldsymbol{\beta}) = \sum_{i=1}^{N} \mathbf{X}_i^T\mathbf{V}_i^{-1}(\mathbf{y}_i - \mathbf{X}_i\boldsymbol{\beta}) = \mathbf{0} \qquad (11.3)$$

where l is the log-likelihood function. The solution is

$$\widehat{\boldsymbol{\beta}} = (\mathbf{X}^T\mathbf{V}^{-1}\mathbf{X})^{-1}\mathbf{X}^T\mathbf{V}^{-1}\mathbf{y} = (\sum_{i=1}^{N} \mathbf{X}_i^T\mathbf{V}_i^{-1}\mathbf{X}_i)^{-1}(\sum_{i=1}^{N} \mathbf{X}_i^T\mathbf{V}_i^{-1}\mathbf{y}_i) \qquad (11.4)$$

with

$$\text{var}(\widehat{\boldsymbol{\beta}}) = (\mathbf{X}^T\mathbf{V}^{-1}\mathbf{X})^{-1} = (\sum_{i=1}^{N} \mathbf{X}_i^T\mathbf{V}_i^{-1}\mathbf{X}_i)^{-1} \qquad (11.5)$$

and $\widehat{\boldsymbol{\beta}}$ is asymptotically Normal (see Chapter 6).

In practice, $\mathbf{V}$ is usually not known and has to be estimated from the data by an iterative process. This involves starting with an initial $\mathbf{V}$ (for instance the identity matrix), calculating an estimate $\widehat{\boldsymbol{\beta}}$ and hence the linear predictors $\widehat{\boldsymbol{\mu}} = \mathbf{X}\widehat{\boldsymbol{\beta}}$ and the residuals $\mathbf{r} = \mathbf{y} - \widehat{\boldsymbol{\mu}}$. The variances and covariances of the residuals are used to calculate $\widehat{\mathbf{V}}$ which in turn is used in (11.5) to obtain a new estimate $\widehat{\boldsymbol{\beta}}$. The process alternates between estimating $\widehat{\boldsymbol{\beta}}$ and estimating $\widehat{\mathbf{V}}$ until convergence is achieved.

If the estimate $\widehat{\mathbf{V}}$ is substituted for $\mathbf{V}$ in equation (11.6), the variance of $\widehat{\boldsymbol{\beta}}$ is likely to be underestimated. Therefore a preferable alternative is

$$\mathbf{V}_s(\widehat{\boldsymbol{\beta}}) = \mathfrak{I}^{-1}\mathbf{C}\mathfrak{I}^{-1}$$

where

$$\mathfrak{I} = \mathbf{X}^T\widehat{\mathbf{V}}^{-1}\mathbf{X} = \sum_{i=1}^{N} \mathbf{X}_i^T\widehat{\mathbf{V}}_i^{-1}\mathbf{X}_i$$

and

$$\mathbf{C} = \sum_{i=1}^{N} \mathbf{X}_i^T\widehat{\mathbf{V}}_i^{-1}(\mathbf{y}_i - \mathbf{X}_i\widehat{\boldsymbol{\beta}})(\mathbf{y}_i - \mathbf{X}_i\widehat{\boldsymbol{\beta}})^T\widehat{\mathbf{V}}_i^{-1}\mathbf{X}_i$$

where $\widehat{\mathbf{V}}_i$ denotes the ith sub-matrix of $\widehat{\mathbf{V}}$. $\mathbf{V}_s(\boldsymbol{\beta})$ is called the **information sandwich estimator**, because $\mathfrak{I}$ is the information matrix (see Chapter 5). It is also sometimes called the **Huber estimator**. It is a consistent estimator of $\text{var}(\widehat{\boldsymbol{\beta}})$ when $\mathbf{V}$ is not known and it is robust to mis-specification of $\mathbf{V}$.

There are several commonly used forms for the matrix $\mathbf{V}_i$.

1. All the off-diagonal elements are equal so that

$$\mathbf{V}_i = \sigma^2 \begin{bmatrix} 1 & \rho & \cdots & \rho \\ \rho & 1 & & \rho \\ \vdots & & \ddots & \vdots \\ \rho & \rho & \cdots & 1 \end{bmatrix}. \tag{11.6}$$

   This is appropriate for clustered data where it is plausible that all measurements are equally correlated, for example, for elements within the same primary sampling unit such as people living in the same area. The term ρ is called the **intra-class correlation coefficient**. The **equicorrelation** matrix in (11.7) is called **exchangeable** or **spherical**. If the off-diagonal term ρ can be written in the form $\sigma_a^2/(\sigma_a^2+\sigma_b^2)$, the matrix is said to have **compound symmetry**.

2. The off-diagonal terms decrease with 'distance' between observations; for example, if all the vectors $\mathbf{y}_i$ have the same length n and

$$\mathbf{V}_i = \sigma^2 \begin{bmatrix} 1 & \rho_{12} & \cdots & \rho_{1n} \\ \rho_{21} & 1 & & \rho_{2n} \\ \vdots & & \ddots & \vdots \\ \rho_{n1} & \rho_{n2} & \cdots & 1 \end{bmatrix} \tag{11.7}$$

   where ρ_{jk} depends on the 'distance' between observations j and k. Examples include $\rho_{jk} = |t_j - t_k|$ for measurements at times t_j and t_k, or $\rho_{jk} = \exp(-|j-k|)$. One commonly used form is the first order **autoregressive model** with $\rho^{|j-k|}$ where $|\rho| < 1$ so that

$$\mathbf{V}_i = \sigma^2 \begin{bmatrix} 1 & \rho & \rho^2 & \cdots & \rho^{n-1} \\ \rho & 1 & \rho & & \rho^{n-2} \\ \rho^2 & \rho & 1 & & \vdots \\ \vdots & & & \ddots & \\ \rho^{n-1} & \cdots & & \rho & 1 \end{bmatrix}. \tag{11.8}$$

3. All the correlation terms may be different

$$\mathbf{V}_i = \sigma^2 \begin{bmatrix} 1 & \rho_{12} & \cdots & \rho_{1n} \\ \rho_{21} & 1 & & \rho_{2n} \\ \vdots & & \ddots & \vdots \\ \rho_{n1} & \rho_{n2} & \cdots & 1 \end{bmatrix}.$$

   This **unstructured correlation matrix** involves no assumptions about correlations between measurements but all the vectors $\mathbf{y}_i$ must be the same length n. It is only practical to use this form when the matrix $\mathbf{V}_i$ is not large relative to the number of subjects because the number, $n(n-1)/2$, of nuisance parameters ρ_{jk} may be excessive and may lead to convergence problems in the iterative estimation process.

The term **repeated measures analysis of variance** is often used when the data are assumed to be Normally distributed. The calculations can be performed using most general purpose statistical software although, sometimes, the correlation structure is assumed to be either spherical or unstructured and correlations which are functions of the times between measurements cannot be modelled. Some programs treat repeated measures as a special case of multivariate data – for example, by not distinguishing between heights of children in the same class (i.e., clustered data), heights of children when they are measured at different ages (i.e., longitudinal data), and heights, weights and girths of children (multivariate data). This is especially inappropriate for longitudinal data in which the time order of the observations matters. The **multivariate approach** to analyzing Normally distributed repeated measures data is explained in detail by Hand and Crowder (1996), while the inappropriateness of these methods for longitudinal data is illustrated by Senn et al. (2000).

11.4 Repeated measures models for non-Normal data

The score equations for Normal models (11.4) can be generalized to other distributions using ideas from Chapter 4. For the generalized linear model

$$\mathrm{E}(Y_i) = \mu_i, \quad g(\mu_i) = \mathbf{x}_i^T \boldsymbol{\beta} = \eta_i$$

for independent random variables $Y_1, Y_2, \ldots, Y_N$ with a distribution from the exponential family, the scores given by equation (4.18) are

$$U_j = \sum_{i=1}^{N} \frac{(y_i - \mu_i)}{\mathrm{var}(Y_i)} x_{ij} \left(\frac{\partial \mu_i}{\partial \eta_i} \right)$$

for parameters $\beta_j, j = 1, \ldots, p$. The last two terms come from

$$\frac{\partial \mu_i}{\partial \beta_j} = \frac{\partial \mu_i}{\partial \eta_i} . \frac{\partial \eta_i}{\partial \beta_j} = \frac{\partial \mu_i}{\partial \eta_i} x_{ij}.$$

Therefore the score equations for the generalized model (with independent responses $Y_i, i = 1, \ldots, N$) can be written as

$$U_j = \sum_{i=1}^{N} \frac{(y_i - \mu_i)}{\mathrm{var}(Y_i)} \frac{\partial \mu_i}{\partial \beta_j} = 0, \qquad j = 1, \ldots, p. \tag{11.9}$$

For repeated measures, let $\mathbf{y}_i$ denote the vector of responses for subject i with $\mathrm{E}(\mathbf{y}_i) = \boldsymbol{\mu}_i$, $g(\boldsymbol{\mu}_i) = \mathbf{X}_i^T \boldsymbol{\beta}$ and let $\mathbf{D}_i$ be the matrix of derivatives $\partial \boldsymbol{\mu}_i / \partial \beta_j$. To simplify the notation, assume that all the subjects have the same number of measurements n.

The **generalized estimating equations** (GEE's) analogous to equations (11.10) are

$$\mathbf{U} = \sum_{i=1}^{N} \mathbf{D}_i^T \mathbf{V}_i^{-1} (\mathbf{y}_i - \boldsymbol{\mu}_i) = \mathbf{0} \tag{11.10}$$

These are also called the **quasi-score equations**. The matrix $\mathbf{V}_i$ can be written as

$$\mathbf{V}_i = \mathbf{A}_i^{\frac{1}{2}}\mathbf{R}_i\mathbf{A}_i^{\frac{1}{2}}\phi$$

where $\mathbf{A}_i$ is the diagonal matrix with elements $\text{var}(y_{ik})$, $\mathbf{R}_i$ is the correlation matrix for $\mathbf{y}_i$ and ϕ is a constant to allow for overdispersion.

Liang and Zeger (1986) showed that if the correlation matrices $\mathbf{R}_i$ are correctly specified, the estimator $\widehat{\boldsymbol{\beta}}$ is consistent and asymptotically Normal. Furthermore, $\widehat{\boldsymbol{\beta}}$ is fairly robust against mis-specification of $\mathbf{R}_i$. They used the term **working correlation matrix** for $\mathbf{R}_i$ and suggested that knowledge of the study design and results from exploratory analyses should be used to select a plausible form. Preferably, $\mathbf{R}_i$ should depend on only a small number of parameters, using assumptions such as equicorrelation or autoregressive correlation (see Section 11.3 above).

The GEE's given by equation (11.11) are used iteratively. Starting with $\mathbf{R}_i$ as the identity matrix and $\phi = 1$, the parameters $\boldsymbol{\beta}$ are estimated by solving equations (11.11). The estimates are used to calculate fitted values $\widehat{\boldsymbol{\mu}}_i = g^{-1}(\mathbf{X}_i^T\boldsymbol{\beta})$ and hence the residuals $\mathbf{y}_i - \widehat{\boldsymbol{\mu}}_i$. These are used to estimate the parameters of $\mathbf{A}_i$, $\mathbf{R}_i$ and ϕ. Then (11.11) is solved again to obtain improved estimates $\widehat{\boldsymbol{\beta}}$, and so on, until convergence is achieved.

Software for solving GEE's is now available in several commercially available software and free-ware programs. While the concepts underlying GEE's are relatively simple there are a number of complications that occur in practice. For example, for binary data, correlation is not a natural measure of association and alternative measures using odds ratios have been proposed (Lipsitz, Laird and Harrington, 1991).

For GEE's it is even more important to use a sandwich estimator for $\text{var}(\widehat{\boldsymbol{\beta}})$ than for the Normal case (see Section 11.3). This is given by

$$\mathbf{V}_s(\widehat{\boldsymbol{\beta}}) = \mathfrak{I}^{-1}\mathbf{C}\mathfrak{I}^{-1}$$

where

$$\mathfrak{I} = \sum_{i=1}^{N} \mathbf{D}_i^T\widehat{\mathbf{V}}_i^{-1}\mathbf{D}_i$$

is the information matrix and

$$\mathbf{C} = \sum_{i=1}^{N} \mathbf{D}_i^T\widehat{\mathbf{V}}_i^{-1}(\mathbf{y}_i - \widehat{\boldsymbol{\mu}}_i)(\mathbf{y}_i - \widehat{\boldsymbol{\mu}}_i)^T\widehat{\mathbf{V}}_i^{-1}\mathbf{D}_i.$$

Then asymptotically, $\widehat{\boldsymbol{\beta}}$ has the distribution $\mathbf{N}\left(\boldsymbol{\beta}, \mathbf{V}_s(\widehat{\boldsymbol{\beta}})\right)$ and inferences can be made using Wald statistics.

11.5 Multilevel models

An alternative approach for analyzing repeated measures data is to use hierarchical models based on the study design. Consider a survey conducted using cluster randomized sampling. Let Y_{jk} denote the response of the kth subject in the jth cluster. For example, suppose Y_{jk} is the income of the kth randomly selected household in council area j, where council areas, the primary sampling units, are chosen randomly from all councils within a country or state. If the goal is to estimate the average household income μ, then a suitable model might be

$$Y_{jk} = \mu + a_j + e_{jk} \tag{11.11}$$

where a_j is the effect of area j and e_{jk} is the random error term. As areas were randomly selected and the area effects are not of primary interest, the terms a_j can be defined as independent, identically distributed random variables with $a_j \sim \mathrm{N}(0, \sigma_a^2)$. Similarly, the terms e_{jk} are independently, identically distributed random variables $e_{jk} \sim \mathrm{N}(0, \sigma_e^2)$ and the a_j's and e_{jk}'s are independent. In this case

$$\mathrm{E}(Y_{jk}) = \mu,$$

$$\mathrm{var}(Y_{jk}) = \mathrm{E}\left[(Y_{jk} - \mu)^2\right] = \mathrm{E}\left[(a_j + e_{jk})^2\right] = \sigma_a^2 + \sigma_e^2,$$

$$\mathrm{cov}(Y_{jk}, Y_{jm}) = \mathrm{E}\left[(a_j + e_{jk})(a_j + e_{jm})\right] = \sigma_a^2$$

for households in the same area, and

$$\mathrm{cov}(Y_{jk}, Y_{lm}) = \mathrm{E}\left[(a_j + e_{jk})(a_l + e_{lm})\right] = 0$$

for households in different areas. If $\mathbf{y}_j$ is the vector of responses for households in area j then the variance-covariance matrix for $\mathbf{y}_j$ is

$$\begin{aligned}
\mathbf{V}_j &= \begin{bmatrix} \sigma_a^2+\sigma_e^2 & \sigma_a^2 & \sigma_a^2 & \cdots & \sigma_a^2 \\ \sigma_a^2 & \sigma_a^2+\sigma_e^2 & \sigma_a^2 & & \sigma_a^2 \\ \sigma_a^2 & \sigma_a^2 & \sigma_a^2+\sigma_e^2 & & \\ \vdots & & & \ddots & \\ \sigma_a^2 & & & \sigma_a^2 & \sigma_a^2+\sigma_e^2 \end{bmatrix} \\
&= \sigma_a^2+\sigma_e^2 \begin{bmatrix} 1 & \rho & \rho & \cdots & \rho \\ \rho & 1 & \rho & & \rho \\ \rho & \rho & 1 & & \\ \vdots & & & \ddots & \\ \rho & & & \rho & 1 \end{bmatrix}
\end{aligned}$$

where $\rho = \sigma_a^2/(\sigma_a^2+\sigma_e^2)$ is the intra-class correlation coefficient. In this case, ρ is the intra-cluster coefficient and it describes the proportion of the total variance due to within-cluster variance. If the responses within a cluster are much more alike than responses from different clusters, then σ_e^2 is much smaller than σ_a^2 so ρ will be near unity; thus ρ is a relative measure of the within-cluster similarity. The matrix $\mathbf{V}_j$ is the same as (11.7), the equicorrelation matrix.

In model (11.12), the parameter μ is a **fixed effect** and the term a_j is a

random effect.This is an example of a **mixed model** with both fixed and random effects. The parameters of interest are μ, σ_a^2 and σ_e^2 (and hence ρ).

As another example, consider longitudinal data in which Y_{jk} is the measurement at time t_k on subject j who was selected at random from the population of interest. A linear model for this situation is

$$Y_{jk} = \beta_0 + a_j + (\beta_1 + b_j)t_k + e_{jk} \tag{11.12}$$

where β_0 and β_1 are the intercept and slope parameters for the population, a_j and b_j are the differences from these parameters specific to subject j, t_k denotes the time of the kth measurement and e_{jk} is the random error term. The terms a_j, b_j and e_{jk} may be considered as random variables with $a_j \sim$N$(0, \sigma_a^2)$, $b_j \sim$N$(0, \sigma_b^2)$, $e_{jk} \sim$N$(0, \sigma_e^2)$ and they are all assumed to be independent. For this model

$$\text{E}(Y_{jk}) = \beta_0 + \beta_1 t_k,$$

$$\text{var}(Y_{jk}) = \text{var}(a_j) + t_k^2\text{var}(b_j) + \text{var}(e_{jk}) = \sigma_a^2 + t_k^2\sigma_b^2 + \sigma_e^2,$$

$$\text{cov}(Y_{jk}, Y_{jm}) = \sigma_a^2 + t_k t_m \sigma_b^2$$

for measurements on the same subject, and

$$\text{cov}(Y_{jk}, Y_{lm}) = 0$$

for measurements on different subjects. Therefore the variance-covariance matrix for subject j is of the form shown in (11.8) with terms dependent on t_k and t_m. In model (11.13), β_0 and β_1 are fixed effects, a_j and b_j are random effects and we want to estimate $\beta_0, \beta_1, \sigma_a^2, \sigma_b^2$ and σ_e^2.

In general, mixed models for Normal responses can be written in the form

$$\mathbf{y} = \mathbf{X}\boldsymbol{\beta} + \mathbf{Zu} + \mathbf{e} \tag{11.13}$$

where $\boldsymbol{\beta}$ are the fixed effects, and $\mathbf{u}$ and $\mathbf{e}$ are random effects. The matrices $\mathbf{X}$ and $\mathbf{Z}$ are design matrices. Both $\mathbf{u}$ and $\mathbf{e}$ are assumed to be Normally distributed. E$(\mathbf{y}) = \mathbf{X}\boldsymbol{\beta}$ summarizes the non-random component of the model. $\mathbf{Zu}$ describes the between-subjects random effects and $\mathbf{e}$ the within-subjects random effects. If $\mathbf{G}$ and $\mathbf{R}$ denote the variance-covariance matrices for $\mathbf{u}$ and $\mathbf{e}$ respectively, then the variance-covariance matrix for $\mathbf{y}$ is

$$\mathbf{V}(\mathbf{y}) = \mathbf{Z}\mathbf{G}^T\mathbf{Z} + \mathbf{R}. \tag{11.14}$$

The parameters of interest are the elements of $\boldsymbol{\beta}$ and the variance and covariance elements in $\mathbf{G}$ and $\mathbf{R}$. For Normal models these can be estimated using the methods of maximum likelihood or residual maximum likelihood (REML). Computational procedures are available in many general purpose statistical programs and more specialized software such as **MLn** (Rabash et al., 1998; Bryk and Raudenbush, 1992). Good descriptions of the use of linear mixed models (especially using the software **SAS**) are given by Verbeke and Molenberghs (1997) and Littell et al. (2000). The books by Longford (1993) and Goldstein (1995) provide detailed descriptions of multilevel, mixed or random coefficient models, predominantly for Normal data.

Mixed models for non-Normal data are less readily implemented although they were first described by Zeger, Liang and Albert (1988) and have been the subject of much research; see, for example, Lee and Nelder (1996). The models are specified as follows

$$\mathrm{E}(\mathbf{y}|\mathbf{u}) = \boldsymbol{\mu}, \qquad \mathrm{var}(\mathbf{y}|\mathbf{u}) = \phi\mathbf{V}(\boldsymbol{\mu}), \qquad g(\boldsymbol{\mu}) = \mathbf{X}\boldsymbol{\beta} + \mathbf{Z}\mathbf{u}$$

where the random coefficients $\mathbf{u}$ have some distribution $f(\mathbf{u})$ and the conditional distribution of $\mathbf{y}$ given $\mathbf{u}$, written as $\mathbf{y}|\mathbf{u}$, follows the usual properties for a generalized linear model with link function g. The unconditional mean and variance-covariance for $\mathbf{y}$ can, in principle, be obtained by integrating over the distribution of $\mathbf{u}$. To make the calculations more tractable, it is common to use conjugate distributions; for example, Normal for $\mathbf{y}|\mathbf{u}$ and Normal for $\mathbf{u}$; Poisson for $\mathbf{y}|\mathbf{u}$ and gamma for $\mathbf{u}$; binomial for $\mathbf{y}|\mathbf{u}$ and beta for $\mathbf{u}$; or binomial for $\mathbf{y}|\mathbf{u}$ and Normal for $\mathbf{u}$. Some software, for example, MLn and Stata can be used to fit mixed or multilevel generalized linear models.

11.6 Stroke example continued

The results of the exploratory analyses and fitting GEE's and mixed models with different intercepts and slopes to the stroke recovery data are shown in Table 11.7. The models were fitted using Stata. Sandwich estimates of the standard errors were calculated for all the GEE models.

Fitting a GEE, assuming independence between observations for the same subject, is the same as the naive or pooled analysis in Table 11.3. The estimate of σ_e is 20.96 (this is the square root of the deviance divided by the degrees of freedom $192 - 6 = 186$). These results suggest that neither intercepts nor slopes differ between groups as the estimates of differences from $\widehat{\alpha}_1$ and $\widehat{\beta}_1$ are small relative to their standard errors.

The data reduction approach which uses the estimated intercepts and slopes for every subject as the data for comparisons of group effects produces the same point estimates but different standard errors. From Tables 11.5 and 11.6, the standard deviations are 21.42 for the intercepts and 3.056 for the slopes and the data do not support hypotheses of differences between the groups.

The GEE analysis, assuming equal correlation among the observations in different weeks, produced the same estimates for the intercept and slope parameters but different standard errors (larger for the intercepts and smaller for the slopes). The estimate of the common correlation coefficient, $\widehat{\rho} = 0.812$, is about the average of the values in Table 11.2 but the assumption of equal correlation is not very plausible. The estimate of σ_e is 20.96, the same as for the models based on independence.

In view of the pattern of correlation coefficients in Table 11.2 an autoregressive model of order 1, AR(1), shown in equation (11.9) seems plausible. In this case, the estimates for ρ and σ_e are 0.964 and 21.08 respectively. The estimates of intercepts and slopes, and their standard errors, differ from the previous

Table 11.7 *Comparison of analyses of the stroke recovery data using various different models.*

	Intercept estimates		
	$\widehat{\alpha}_1$ (s.e.)	$\widehat{\alpha}_2 - \widehat{\alpha}_1$ (s.e.)	$\widehat{\alpha}_3 - \widehat{\alpha}_1$ (s.e.)
Pooled	29.821 (5.774)	3.348 (8.166)	-0.022 (8.166)
Data reduction	29.821 (5.772)	3.348 (10.709)	-0.018 (10.709)
GEE, independent	29.821 (5.774)	3.348 (8.166)	-0.022 (8.166)
GEE, equicorrelated	29.821 (7.131)	3.348 (10.085)	-0.022 (10.085)
GEE, AR(1)	33.538 (7.719)	-0.342 (10.916)	-6.474 (10.916)
GEE, unstructured	30.588 (7.462)	2.319 (10.552)	-1.195 (10.552)
Random effects	29.821 (7.047)	3.348 (9.966)	-0.022 (9.966)
	Slope estimates		
	$\widehat{\beta}_1$ (s.e.)	$\widehat{\beta}_2 - \widehat{\beta}_1$ (s.e.)	$\widehat{\beta}_3 - \widehat{\beta}_1$ (s.e.)
Pooled	6.324 (1.143)	-1.994 (1.617)	-2.686 (1.617)
Data reduction	6.324 (1.080)	-1.994 (1.528)	-2.686 (1.528)
GEE, independent	6.324 (1.143)	-1.994 (1.617)	-2.686 (1.617)
GEE, equicorrelated	6.324 (0.496)	-1.994 (0.701)	-2.686 (0.701)
GEE, AR(1)	6.073 (0.714)	-2.142 (1.009)	-2.686 (1.009)
GEE, unstructured	6.926 (0.941)	-3.214 (1.331)	-2.686 (1.331)
Random effects	6.324 (0.463)	-1.994 (0.655)	-2.686 (0.655)

models. Wald statistics for the differences in slope support the hypothesis that the patients in group A improved significantly faster than patients in the other two groups.

The GEE model with an unstructured correlation matrix involved fitting 28 (8×7/2) correlation parameters. The estimate of σ_e was 21.21. While the point estimates differ from those for the other GEE models with correlation, the conclusion that the slopes differ significantly is the same.

The final model fitted was the mixed model (11.13) estimated by the method of maximum likelihood. The point estimates and standard errors for the fixed parameters were similar to those from the GEE model with the equicorrelated matrix. This is not surprising as the estimated intra-class correlation coefficient is $\widehat{\rho} = 0.831$.

This example illustrates both the importance of taking into account the correlation between repeated measures and the robustness of the results regardless of how the correlation is modelled. Without considering the correlation it was not possible to detect the statistically significantly better outcomes for patients in group A.

11.7 Comments

Exploratory analyses for repeated measures data should follow the main steps outlined in Section 11.2. For longitudinal data these include plotting the time course for individual subjects or groups of subjects, and using an appropriate form of data reduction to produce summary statistics that can be examined to identify patterns for the population overall or for sub-samples. For clustered data it is worthwhile to calculate summary statistics at each level of a multilevel model to examine both the main effects and the variability.

Missing data can present problems. With suitable software it may be possible to perform calculations on unbalanced data (e.g., different numbers of observations per subject) but this is dangerous without careful consideration of why data are missing. Occasionally they may be **missing completely at random**, unrelated to observed responses or any covariates (Little and Rubin, 1987). In this case, the results should be unbiased. More commonly, there are reasons why the data are missing. For example, in a longitudinal study of treatment some subjects may have become too ill to continue in the study, or in a clustered survey, outlying areas may have been omitted due to lack of resources. In these cases results based on the available data will be biased. Diggle, Liang and Zeger (1994), Diggle and Kenward (1994) and Trozel, Harrington and Lipsitz (1998) discuss the problem in more detail and provide some suggestions about how adjustments may be made in some situations.

Unbalanced data and longitudinal data in which the observations are not equally spaced or do not all occur at the planned times can be accommodated in mixed models and generalized estimating equations; for example, see Cnaan et al. (1997), Burton et al. (1998) and Carlin et al.(1999).

Inference for models fitted by GEE's is best undertaken using Wald statistics with a robust sandwich estimator for the variance. The optimal choice of the correlation matrix is not critical because the estimator is robust with respect to the choice of working correlation matrix, but a poor choice can reduce the efficiency of the estimator. In practice, the choice may be affected by the number of correlation parameters to be estimated; for example, use of a large unstructured correlation matrix may produce unstable estimates or the calculations may not converge. Selection of the correlation matrix can be done by fitting models with alternative covariance structures and comparing the **Akaike information criterion**, which is a function of the log-likelihood function adjusted for the number of covariance parameters (Cnaan et al., 1997). Model checking can be carried out with the usual range of residual plots.

For multilevel data, nested models can be compared using likelihood ratio statistics. Residuals used for checking the model assumptions need to be standardized or 'shrunk', to apportion the variance appropriately at each level of the model (Goldstein, 1995). If the primary interest is in the random effects then Bayesian methods for analyzing the data, for example, using BUGS, may be more appropriate than the frequentist approach adopted here (Best and Speigelhalter, 1996).

Table 11.8 *Measurements of left ventricular volume and parallel conductance volume on five dogs under eight different load conditions: data from Boltwood et al. (1989).*

		Conditions							
Dog		1	2	3	4	5	6	7	8
1	y	81.7	84.3	72.8	71.7	76.7	75.8	77.3	86.3
	x	54.3	62.0	62.3	47.3	53.6	38.0	54.2	54.0
2	y	105.0	113.6	108.7	83.9	89.0	86.1	88.7	117.6
	x	81.5	80.8	74.5	71.9	79.5	73.0	74.7	88.6
3	y	95.5	95.7	84.0	85.8	98.8	106.2	106.4	115.0
	x	65.0	68.3	67.9	61.0	66.0	81.8	71.4	96.0
4	y	113.1	116.5	100.8	101.5	120.8	95.0	91.9	94.0
	x	87.5	93.6	70.4	66.1	101.4	57.0	82.5	80.9
5	y	99.5	99.2	106.1	85.2	106.3	84.6	92.1	101.2
	x	79.4	82.5	87.9	66.4	68.4	59.5	58.5	69.2

11.8 Exercises

11.1 The measurement of left ventricular volume of the heart is important for studies of cardiac physiology and clinical management of patients with heart disease. An indirect way of measuring the volume, y, involves a measurement called parallel conductance volume, x. Boltwood et al. (1989) found an approximately linear association between y and x in a study of dogs under various 'load' conditions. The results, reported by Glantz and Slinker (1990), are shown in Table 11.8.

(a) Conduct an exploratory analysis of these data.

(b) Let (Y_{jk}, x_{jk}) denote the kth measurement on dog j, $(j = 1, \ldots, 5; k = 1, \ldots, 8)$. Fit the linear model

$$\mathrm{E}(Y_{jk}) = \mu = \alpha + \beta x_{jk}, \qquad Y \sim N(\mu, \sigma^2),$$

assuming the random variables Y_{jk} are independent (i.e., ignoring the repeated measures on the same dogs). Compare the estimates of the intercept α and slope β and their standard errors from this pooled analysis with the results you obtain using a data reduction approach.

(c) Fit a suitable random effects model.

(d) Fit a longitudinal model using a GEE.

(e) Compare the results you obtain from each approach. Which method(s) do you think are most appropriate? Why?

11.2 Suppose that (Y_{jk}, x_{jk}) are observations on the kth subject in cluster k (with $j = 1, \ldots, J; k = 1, \ldots, K$) and we want to fit a 'regression through the origin' model

$$\mathrm{E}(Y_{jk}) = \beta x_{jk}$$

where the variance-covariance matrix for Y's in the same cluster is

$$\mathbf{V}_j = \sigma^2 \begin{bmatrix} 1 & \rho & \cdots & \rho \\ \rho & 1 & & \rho \\ \vdots & & \ddots & \vdots \\ \rho & \rho & \cdots & 1 \end{bmatrix}$$

and Y's in different clusters are independent.

(a) From Section 11.3, if the Y's are Normally distributed then

$$\widehat{\boldsymbol{\beta}} = (\sum_{j=1}^{J} \mathbf{x}_j^T \mathbf{V}_j^{-1} \mathbf{x}_j)^{-1} (\sum_{j=1}^{J} \mathbf{x}_j^T \mathbf{V}_j^{-1} \mathbf{y}_j) \text{ with } \operatorname{var}(\widehat{\boldsymbol{\beta}}) = (\sum_{j=1}^{J} \mathbf{x}_j^T \mathbf{V}_j^{-1} \mathbf{x}_j)^{-1}$$

where $\mathbf{x}_j^T = [x_{j1}, \ldots, x_{jK}]$. Deduce that the estimate b of β is unbiased.

(b) As

$$\mathbf{V}_j^{-1} = c \begin{bmatrix} 1 & \phi & \cdots & \phi \\ \phi & 1 & & \phi \\ \vdots & & \ddots & \vdots \\ \phi & \phi & \cdots & 1 \end{bmatrix}$$

$$\text{where } c = \frac{1}{\sigma^2[1 + (K-1)\phi\rho]} \quad \text{and } \phi = \frac{-\rho}{1 + (K-2)\rho}$$

show that

$$\operatorname{var}(b) = \frac{\sigma^2[1 + (K-1)\phi\rho]}{\sum_j \{\sum_k x_{jk}^2 + \phi[(\sum_k x_{jk})^2 - \sum_k x_{jk}^2]\}}.$$

(c) If the clustering is ignored, show that the estimate b^* of β has $\operatorname{var}(b^*) = \sigma^2 / \sum_j \sum_k x_{jk}^2$.

(d) If $\rho = 0$, show that $\operatorname{var}(b) = \operatorname{var}(b^*)$ as expected if there is no correlation within clusters.

(e) If $\rho = 1$, $\mathbf{V}_j/\sigma^2$ is a matrix of one's so the inverse doesn't exist. But the case of maximum correlation is equivalent to having just one element per cluster. If $K = 1$, show that $\operatorname{var}(b) = \operatorname{var}(b^*)$, in this situation.

(f) If the study is designed so that $\sum_k x_{jk} = 0$ and $\sum_k x_{jk}^2$ is the same for all clusters, let $W = \sum_j \sum_k x_{jk}^2$ and show that

$$\operatorname{var}(b) = \frac{\sigma^2[1 + (K-1)\phi\rho]}{W(1-\phi)}.$$

(g) With this notation $\operatorname{var}(b^*) = \sigma^2/W$, hence show that

$$\frac{\operatorname{var}(b)}{\operatorname{var}(b^*)} = \frac{[1 + (K-1)\phi\rho]}{1-\phi} = 1 - \rho.$$

Deduce the effect on the estimated standard error of the slope estimate for this model if the clustering is ignored.

Table 11.9 *Numbers of ears clear of acute otitis media at 14 days, tabulated by antibiotic treatment and age of the child: data from Rosner(1989).*

	CEF				AMO			
	Number clear				Number clear			
Age	0	1	2	Total	0	1	2	Total
< 2	8	2	8	18	11	2	2	15
$2 - 5$	6	6	10	22	3	1	5	9
≥ 6	0	1	3	4	1	0	6	7
Total	14	9	21	44	15	3	13	31

11.3 Data on the ears or eyes of subjects are a classical example of clustering – the ears or eyes of the same subject are unlikely to be independent. The data in Table 11.9 are the responses to two treatments coded CEF and AMO of children who had acute otitis media in both ears (data from Rosner, 1989).

(a) Conduct an exploratory analysis to compare the effects of treatment and age of the child on the success of the treatments, ignoring the clustering within each child.

(b) Let Y_{ijkl} denote the response of the lth ear of the kth child in the treatment group j and age group i. The Y'_{ijkl}'s are binary variables with possible values of 1 denoting 'cured' and 0 denoting 'not cured'. A possible model is

$$\log\text{it}\left(\frac{\pi_{ijkl}}{1-\pi_{ijkl}}\right) = \beta_0 + \beta_1\text{age} + \beta_2\text{treatment} + b_k$$

where b_k denotes the random effect for the kth child and β_0, β_1 and β_2 are fixed parameters. Fit this model (and possibly other related models) to compare the two treatments. How well do the models fit? What do you conclude about the treatments?

(c) An alternative approach, similar to the one proposed by Rosner, is to use nominal logistic regression with response categories 0, 1 or 2 cured ears for each child. Fit a model of this type and compare the results with those obtained in (b). Which approach is preferable considering the assumptions made, ease of computation and ease of interpretation?

Software

Genstat

Numerical Algorithms Group, NAG Ltd, Wilkinson House, Jordan Hill Road, Oxford, OX2 8DR, United Kingdom

http://www.nag.co.uk/

Glim

Numerical Algorithms Group, NAG Ltd, Wilkinson House, Jordan Hill Road, Oxford, OX2 8DR, United Kingdom

http://www.nag.co.uk/

Minitab

Minitab Inc., 3081 Enterprise Drive, State College, PA 16801-3008, U.S.A.

http://www.minitab.com

MLwiN

Multilevel Models Project, Institute of Education, 20 Bedford Way, London, WC1H OAL, United Kingdom

http://multilevel.ioe.ac.uk/index.html

Stata

Stata Corporation, 4905 Lakeway Drive, College Station, Texas 77845, U.S.A.

http://www.stata.com

SAS

SAS Institute Inc., SAS Campus Drive, Cary, NC 27513-2414, U.S.A.

http://www.sas.com

SYSTAT

SPSS Science, 233 S. Wacker Drive, 11th Floor, Chicago, IL 60606-6307, U.S.A.

http://www.spssscience.com/SYSTAT/

S-PLUS

Insightful Corporation, 1700 Westlake Avenue North, Suite 500, Seattle, WA 98109-3044, U.S.A.

http://www.insightful.com/products/splus/splus2000/splusstdintro.html

StatXact and LogXact

Cytel Software Corporation, 675 Massachusetts Avenue, Cambridge, MA 02139, U.S.A.

http://www.cytel.com/index.html

References

Aitkin, M., Anderson, D., Francis, B. and Hinde, J. (1989) *Statistical Modelling in GLIM*, Clarendon Press, Oxford.

Altman, D. G., Machin, D., Bryant, T. N. and Gardner, M. J. (2000) *Statistics with Confidence*, British Medical Journal, London.

Altman, D. G. and Bland, J. M. (1998) Statistical notes: times to event (survival) data. *British Medical Journal*, 317, 468-469.

Agresti, A. (1990) *Categorical Data Analysis*, Wiley, New York.

Agresti, A. (1996) *An Introduction to Categorical Data Analysis*, Wiley, New York.

Ananth, C. V. and Kleinbaum, D. G. (1997) Regression models for ordinal responses: a review of methods and applications. *International Journal of Epidemiology*, 26, 1323-1333.

Andrews, D. F. and Herzberg, A. M. (1985) *Data: A Collection of Problems from Many Fields for the Student and Research Worker*, Springer Verlag, New York.

Aranda-Ordaz, F. J. (1981) On two families of transformations to additivity for binary response data. *Biometrika*, 68, 357-363.

Ashby, D., West, C. R. and Ames, D. (1989) The ordered logistic regression model in psychiatry: rising prevalence of dementia in old people's homes. *Statistics in Medicine*, 8, 1317-1326.

Australian Bureau of Statistics (1998) *Apparent consumption of foodstuffs 1996-97*, publication 4306.0, Australian Bureau of Statistics, Canberra.

Baxter, L. A., Coutts, S. M. and Ross, G. A. F. (1980) Applications of linear models in motor insurance, in *Proceedings of the 21st International Congress of Actuaries*, Zurich, 11-29.

Belsley, D. A., Kuh, E. and Welsch, R. E. (1980). *Regression Diagnostics: Identifying Influential Observations and Sources of Collinearity*, Wiley, New York.

Best, N. G., Speigelhalter, D. J., Thomas, A. and Brayne, C. E. G. (1996) Bayesian analysis of realistically complex models. *Journal of the Royal Statistical Society, Series A*, 159, 323-342.

Birch, M. W. (1963) Maximum likelihood in three-way contingency tables. *Journal of the Royal Statistical Society, Series B*, 25, 220-233.

Bliss, C. I. (1935) The calculation of the dose-mortality curve. *Annals of Applied Biology*, 22, 134-167.

Boltwood, C. M., Appleyard, R. and Glantz, S. A. (1989) Left ventricular volume measurement by conductance catheter in intact dogs: the parallel conductance volume increases with end-systolic volume. *Circulation*, 80, 1360-1377.

Box, G. E. P. and Cox, D. R. (1964). An analysis of transformations. *Journal of the Royal Statistical Society, Series B*, 26, 211-234.

Breslow, N. E. and Day, N. E. (1987) *Statistical Methods in Cancer Research, Volume 2: The Design and Analysis of Cohort Studies*, International Agency for Research on Cancer, Lyon.

Brown, C. C. (1982) On a goodness of fit test for the logistic model based on score statistics. *Communications in Statistics – Theory and Methods*, 11, 1087-1105.

Brown, W., Bryson, L., Byles, J., Dobson, A., Manderson, L., Schofield, M. and Williams, G. (1996) Women's Health Australia: establishment of the Australian Longitudinal Study on Women's Health. *Journal of Women's Health*, 5, 467-472.

Bryk, A. S. and Raudenbush, S. W. (1992). *Hierarchical Linear Models: Applications and Data Analysis Methods*. Sage, Thousand Oaks, California.

Burton, P., Gurrin, L. and Sly, P. (1998) Tutorial in biostatistics – extending the simple linear regression model to account for correlated responses: an introduction to generalized estimating equations and multi-level modelling. *Statistics in Medicine*, 17, 1261-1291.

Cai, Z. and Tsai, C.-L. (1999) Diagnostics for non-linearity in generalized linear models. *Computational Statistics and Data Analysis*, 29, 445-469.

Carlin, J. B., Wolfe, R. Coffey, C. and Patton, G. (1999) Tutorial in biostatistics – analysis of binary outcomes in longitudinal studies using weighted estimating equations and discrete-time survival methods: prevalence and incidence of smoking in an adolescent cohort. *Statistics in Medicine*, 18, 2655-2679.

Charnes, A., Frome, E. L. and Yu, P. L. (1976) The equivalence of generalized least squares and maximum likelihood estimates in the exponential family. *Journal of the American Statistical Association*, 71, 169-171.

Chatterjee, S. and Hadi, A. S. (1986) Influential observations, high leverage points, and outliers in linear regression. *Statistical Science*, 1, 379-416.

Cnaan, A., Laird, N. M. and Slasor, P. (1997) Tutorial in biostatistics – using the generalized linear model to analyze unbalanced repeated measures and longitudinal data. *Statistics in Medicine*, 16, 2349-2380.

Collett, D. (1991) *Modelling Binary Data*, Chapman and Hall, London.

Collett, D. (1994) *Modelling Survival Data in Medical Research*, Chapman and Hall, London.

Cook, R. D. and Weisberg, S. (1999) *Applied Regression including Computing and Graphics*, Wiley, New York.

Cox, D. R. (1972) Regression models and life tables (with discussion). *Journal of the Royal Statistical Society, Series B*, 74, 187-220.

Cox, D. R. and Hinkley, D. V. (1974) *Theoretical Statistics*, Chapman and Hall, London.

Cox, D. R. and Oakes, D. V. (1984) *Analysis of Survival Data*, Chapman and Hall, London.

Cox, D. R. and Snell, E. J. (1968) A general definition of residuals. *Journal of the Royal Statistical Society, Series B*, 30, 248-275.

Cox, D. R. and Snell, E. J. (1981) *Applied Statistics: Principles and Examples*, Chapman and Hall, London.

Cox, D. R. and Snell, E. J. (1989) *Analysis of Binary Data, Second Edition,* Chapman and Hall, London.

Cressie, N. and Read, T. R. C. (1989) Pearson's χ^2 and the loglikelihood ratio statistic G^2: a comparative review. *International Statistical Review*, 57, 19-43.

Crowder, M. J., Kimber, A. C., Smith, R. L. and Sweeting, T. J. (1991) *Statistical Analysis of Reliability Data*, Chapman and Hall, London.

Crowley, J. and Hu, M. (1977) Covariance analysis of heart transplant survival data. *Journal of the American Statistical Association,* 72, 27-36.

Diggle, P. and Kenward, M. G. (1994) Informative drop-out in longitudinal data analysis. *Applied Statistics*, 43, 49-93.

Diggle, P. J., Liang, K.-Y. and Zeger, S. L. (1994) *Analysis of Longitudinal Data*, Oxford University Press, Oxford.

Dobson, A. J. and Stewart, J. (1974) Frequencies of tropical cyclones in the northeastern Australian area. *Australian Meteorological Magazine*, 22, 27-36.

Draper, N. R. and Smith, H. (1998) *Applied Regression Analysis, Third Edition*, Wiley-Interscience, New York.

Duggan, J. M., Dobson, A. J., Johnson, H. and Fahey, P. P. (1986) Peptic ulcer and non-steroidal anti-inflammatory agents. *Gut*, 27, 929-933.

Egger, G., Fisher, G., Piers, S., Bedford, K., Morseau, G., Sabasio, S., Taipim, B., Bani, G., Assan, M. and Mills, P. (1999) Abdominal obesity reduction in Indigenous men. *International Journal of Obesity*, 23, 564-569.

Evans, M., Hastings, N. and Peacock, B. (2000) *Statistical Distributions, Third Edition*, Wiley, New York.

Fahrmeir, L. and Kaufmann, H. (1985) Consistency and asymptotic normality of the maximum likelihood estimator in generalized linear models. *Annals of Statistics*, 13, 342-368.

Feigl, P. and Zelen, M. (1965) Estimation of exponential probabilities with concomitant information. *Biometrics*, 21, 826-838.

Finney, D. J. (1973) *Statistical Methods in Bioassay, Second Edition*, Hafner, New York.

Fleming, T. R. and Harrington, D. P. (1991) *Counting Processes and Survival Analysis*, Wiley, New York.

Fuller, W. A. (1987) *Measurement Error Models*, Wiley, New York.

Gehan, E. A. (1965) A generalized Wilcoxon test for comparing arbitrarily singly-censored samples. *Biometrika*, 52, 203-223.

Glantz, S. A. and Slinker, B. K. (1990) *Primer of Applied Regression and Analysis of Variance*, McGraw Hill, New York.

Goldstein, H. (1995) *Multilevel Statistical Modelling, Second Edition*, Arnold, London.

Graybill, F. A. (1976) *Theory and Application of the Linear Model*, Duxbury, North Scituate, Massachusetts.

Hand, D. and Crowder, M. (1996) *Practical Longitudinal Data Analysis*, Chapman and Hall, London.

Hastie, T. J. and Tibshirani, R. J. (1990) *Generalized Additive Models*, Chapman and Hall, London.

Healy, M. J. R. (1988) *GLIM: An Introduction*, Clarendon Press, Oxford.

Hogg, R. V. and Craig, A. T. (1995) *Introduction to Mathematical Statistics, Fifth Edition*, Prentice Hall, New Jersey.

Holtbrugger, W. and Schumacher, M. (1991) A comparison of regression models for the analysis of ordered categorical data. *Applied Statistics*, 40, 249-259.

Hosmer, D. W. and Lemeshow, S. (1980) Goodness of fit tests for the multiple logistic model. *Communications in Statistics – Theory and Methods*, A9, 1043-1069.

Hosmer, D. W. and Lemeshow, S. (2000) *Applied Logistic Regression, Second Edition*, Wiley, New York.

Jones, R. H. (1987) Serial correlation in unbalanced mixed models. *Bulletin of the International Statistical Institute*, 52, Book 4, 105-122.

Kalbfleisch, J. G. (1985) *Probability and Statistical Inference, Volume 2: Statistical Inference, Second Edition*, Springer-Verlag, New York

Kalbfleisch, J. D. and Prentice, R. L. (1980) *The Statistical Analysis of Failure Time Data*, Wiley, New York.

Kleinbaum, D. G., Kupper, L. L., Muller, K. E. and Nizam, A. (1998) *Applied Regression Analysis and Multivariable Methods, Third Edition*, Duxbury, Pacific Grove, California.

Krzanowski, W. J. (1998) *An Introduction to Statistical Modelling*, Arnold, London.

Lee, E. T. (1992) *Statistical Methods for Survival Data Analysis*, Wiley, New York.

Lee, Y. and Nelder, J. A. (1996) Hierarchical generalized linear models (with discussion). *Journal of the Royal Statistical Society, Series B*, 58, 619-678.

Lewis, T. (1987) Uneven sex ratios in the light brown apple moth: a problem in outlier allocation, in *The Statistical Consultant in Action*, edited by D. J. Hand and B. S. Everitt, Cambridge University Press, Cambridge.

Liang, K.-Y. and Zeger, S. L (1986) Longitudinal data analysis using generalized linear models. *Biometrika*, 73, 13-22.

Lipsitz, S. R., Laird, N. M. and Harrington, D. P. (1991) Generalized estimating equations for correlated binary data: using the odds ratio as a measure of association. *Biometrika*, 78, 153-160.

Littell, R. C., Pendergast, J. and Natarajan, R. (2000) Tutorial in biostatistics – modelling covariance structure in the analysis of repeated measures data. *Statistics in Medicine*, 19, 1793-1819.

Little, R. J. A. and Rubin, D. B. (1987) *Statistical Analysis with Missing Data.* Wiley, New York.

Longford, N. (1993) *Random Coefficient Models.* Oxford University Press, Oxford.

Madsen, M. (1971) Statistical analysis of multiple contingency tables. Two examples. *Scandinavian Journal of Statistics*, 3, 97-106.

McCullagh, P. (1980) Regression models for ordinal data (with discussion). *Journal of the Royal Statistical Society, Series B*, 42, 109-142.

McCullagh, P. and Nelder, J. A. (1989) *Generalized Linear Models, Second Edition*, Chapman and Hall, London.

McFadden, M., Powers, J., Brown, W. and Walker, M. (2000) Vehicle and driver attributes affecting distance from the steering wheel in motor vehicles. *Human Factors*, 42, 676-682.

McKinlay, S. M. (1978) The effect of nonzero second-order interaction on combined estimators of the odds ratio. *Biometrika*, 65, 191-202.

Montgomery, D. C. and Peck, E. A. (1992) *Introduction to Linear Regression Analysis, Second Edition*, Wiley, New York.

National Centre for HIV Epidemiology and Clinical Research (1994) *Australian HIV Surveillance Report*, 10.

Nelder, J. A. and Wedderburn, R. W. M. (1972) Generalized linear models. *Journal of the Royal Statistical Society, Series A*, 135, 370-384.

Neter, J., Kutner, M. H., Nachtsheim, C. J. and Wasserman, W. (1996) *Applied Linear Statistical Models, Fourth Edition*, Irwin, Chicago.

Otake, M. (1979) *Comparison of Time Risks Based on a Multinomial Logistic Response Model in Longitudinal Studies.* Technical Report No. 5, RERF, Hiroshima, Japan.

Pierce, D. A. and Schafer, D. W. (1986) Residuals in generalized linear models. *Journal of the American Statistical Association*, 81, 977-986.

Prigibon, D. (1981) Logistic regression diagnostics. *Annals of Statistics*, 9, 705-724.

Rasbash, J., Woodhouse, G., Goldstein, H., Yang, M. and Plewis, I. (1998) *MLwiN Command Reference.* Institute of Education, London.

Rao, C. R. (1973) *Linear Statistical Inference and Its Applications, Second Edition.* Wiley, New York.

Roberts, G., Martyn, A. L., Dobson, A. J. and McCarthy, W. H. (1981) Tumour thickness and histological type in malignant melanoma in New South Wales, Australia, 1970-76. *Pathology*, 13, 763-770.

Rosner, B. (1989) Multivariate methods for clustered binary data with more than one level of nesting. *Journal of the American Statistical Association*, 84, 373-380.

Sangwan-Norrell, B. S. (1977) Androgenic stimulating factor in the anther and isolated pollen grain culture of *Datura innoxia* Mill. *Journal of Experimental Biology*, 28, 843-852.

Senn, S., Stevens, L. and Chaturvedi, N. (2000) Tutorial in biostatistics – repeated measures in clinical trials: simple strategies for analysis using summary measures. *Statistics in Medicine*, 19, 861-877.

Sinclair, D. F. and Probert, M. E. (1986) A fertilizer response model for a mixed pasture system, in *Pacific Statistical Congress* (Eds I. S. Francis et al.) Elsevier, Amsterdam, 470-474.

Trozel, A. B., Harrington, D. P. and Lipsitz, S. R. (1998) Analysis of longitudinal data with non-ignorable non-monotone missing values. *Applied Statistics*, 47, 425-438.

Verbeke, G. and Molenberghs, G. (1997) *Linear Mixed Models in Practice*. Springer Verlag, New York.

Wei, L. J. (1992) The accelerated failure time model: a useful alternative to the Cox regression model in survival analysis. *Statistics in Medicine*, 11, 1871-1879.

Winer, B. J. (1971) *Statistical Principles in Experimental Design, Second Edition,* McGraw-Hill, New York.

Wood, C. L. (1978) Comparison of linear trends in binomial proportions. *Biometrics*, 34, 496-504.

Zeger, S. L., Liang, K.-Y. and Albert, P. (1988) Models for longitudinal data: a generalized estimating equation approach. *Biometrics*, 44, 1049-60.

Index